Crocologia—A Detailed Study of Saffron, the K

Brill's Studies in Intellectual History

General Editor

Han van Ruler (*Erasmus University, Rotterdam*)

Founded by

Arjo Vanderjagt

VOLUME 319

Brill's Texts and Sources in Intellectual History

General Editor

Leen Spruit (*Radboud University, Nijmegen*)

VOLUME 22

The titles published in this series are listed at *brill.com/btsi*

Crocologia—A Detailed Study of Saffron, the King of Plants

By

Sally Francis
Maria Teresa Ramandi

TUTA SUB AEGIDE PALLAS · 1683 ·

BRILL

LEIDEN | BOSTON

Cover illustration: Picture copyright Sally Francis.

Library of Congress Cataloging-in-Publication Data

Names: Hertodt, Johann Ferdinand, 1647–1714, author. | Francis, Sally, Dr., writer of supplementary textual content, translator. | Ramandi, Maria Teresa, Dr., writer of supplementary textual content, translator.
Title: Crocologia–a detailed study of saffron, the king of plants / by Sally Francis, Maria Teresa Ramandi.
Other titles: Crocologia seu curia croci regis vegetabilium enucleatio. English | Brill's studies in intellectual history ; v. 319. 0920-8607
Description: Leiden ; Boston : Brill, [2020] | Series: Brill's studies in intellectual history, 0920-8607 ; 319. Brill's texts and sources in intellectual history ; v. 22 | Includes bibliographical references and index.
Identifiers: LCCN 2020023177 (print) | LCCN 2020023178 (ebook) | ISBN 9789004423534 (hardback ; alk. paper) | ISBN 9789004435292 (ebook)
Subjects: MESH: Crocus | Plants, Medicinal | Plant Preparations–history | Plant Preparations–therapeutic use
Classification: LCC QK495.I75 (print) | LCC QK495.I75 (ebook) | NLM WZ 290 | DDC 584/.75–dc23
LC record available at https://lccn.loc.gov/2020023177
LC ebook record available at https://lccn.loc.gov/2020023178

Typeface for the Latin, Greek, and Cyrillic scripts: "Brill". See and download: brill.com/brill-typeface.

ISSN 0920-8607
ISBN 978-90-04-42353-4 (hardback)
ISBN 978-90-04-43529-2 (e-book)

This book is printed on acid-free paper and produced in a sustainable manner.

Contents

Acknowledgments

This book is the result of a fruitful collaboration between a botanist saffron-grower and a Latinist. It would have been impossible to achieve our translation of *Crocologia*, and write our accompanying chapters, without the help of the following people. First, I would like to thank Dr Emma Spary, Faculty of History, University of Cambridge for her enthusiastic response to my ideas about *Crocologia*, and for putting me in touch with my co-author, Dr Maria Teresa Ramandi. It has been a pleasure to work with Maria and I pay tribute here to her vast expertise and professionalism.

The following people all provided invaluable information used in our book: Tony Goode, UK *Crocus* National Collection Holder; Dr Bernard Kaar, Wachauer Safran, Austria; Dr Christopher Etheridge, British Herbal Medicine Association; and Francis Treuherz, Homeopath. I thank Britta Tiemann for translating a work by Abl.

I thank Dr Barrie Juniper, Department of Plant Sciences, University of Oxford for his comments on our manuscript. I am very grateful to Amelia Walker, Wellcome Collection, London; Stuart Tiley, Balliol College, Oxford; and Corye Bradbury, ProQuest LLC, USA, for facilitating access to the three images from the original version of *Crocologia*.

Finally I must thank my editor Ivo Romein, for his positive response to our book proposal, Professor Leen Spruit (the series editor) for his comments on the manuscript, and Ivo and his colleagues at Brill for making this book come to fruition.

Sally Francis
Burnham Norton, 27th March 2020

Disclaimer

Figures

Introduction

> Those who are desirous of knowing the pharmaceutical preparation of saffron, and the diseases in the curing of which it was employed, may read Hertodt's *Crocologia*, where the author has collected all the receipts ... for preparing it.[1]

Here we present the first translation into English of *Crocologia*, by Dr Johann Ferdinand Hertodt, originally published in Latin in 1671. This remarkable, unique work focuses entirely on saffron (*Crocus sativus* L.) and its therapeutic uses, effectively presenting us with a saffron-orientated literature review of Western European early modern medicine.[2]

Our introduction is broadly organised into four sections: (1) Hertodt's biography, and a description of *Crocologia*; (2) a contemporary narrative on saffron, its origin, cultivation, adulteration etc., (3) the changing role of saffron in orthodox medicine after *Crocologia*, and modern clinical trial findings, and (4) a conclusion.

1 *Crocologia* and Its Author

1.1 *Johann Ferdinand Hertodt*

Crocologia's author, Johann Ferdinand Hertodt von Todtenfeld was a physician and a naturalist. He wrote on several subjects, including alchemy, botany, geology and medicine. Despite the fact that his works enjoyed a wide circulation within the scientific community of the time, very little is known about is life.

Our author was born to an aristocratic family on 4th February 1647 in the town of Mikulov (then called Nikolsburg) in Southern Moravia.[3] Precise information about his parents is unavailable, although the surname "Hertodt"

1 Johnston, 1797, p. 280.

2 Details of seven later monographs on saffron, published in Latin, French and Italian, are listed by Waring, 1878, p. 389.

3 Moravia was a region of the Holy Roman Empire corresponding more or less to the eastern part of today's Czech Republic. Its capital was Brno.

 | DOI:10.1163/9789004435292_002

seems to come from Hungary and is listed among the historical surnames of the Hungarian nobility.[4] Hertodt studied medicine in Padua,[5] where he graduated in 1666 aged 19, but he was not the only physician in his family. His cousin,[6] Matthias Franz Hertodt von Todtenfeld (1625–1712), practised in Mikulov,[7] held the position of provincial physician (*Landschaftsphysicus*) in Brno and Znojmo,[8] as well as that of personal physician and imperial counsellor to the Emperor Leopold I,[9] and later to the Emperor Charles VI.[10] Matthias Franz Hertodt also authored *Contemplatio, super sanguinis & spiritus, hominum ad atrophiam senilem deflectentium* (The Hague, 1712).[11] Another Hertodt, Johann Phillip (d. 1707; Matthias Franz's brother and, therefore, another cousin of Johann Ferdinand), was a municipal physician (*Stadtphysicus*) in Znojmo (roughly 50 km west of Mikulov) and the writer of *Medicus euporistus* (The Hague, 1711), a work on the twelve universal medications.[12]

In 1668 Johann Ferdinand Hertodt was appointed municipal physician of Brno (approximately 50 km north of Mikulov), and only a few years later, in 1671, Emperor Leopold I nominated him provincial physician of Moravia. The years between his graduation and his appointment as provincial physician of Moravia were fundamental in Hertodt's career. It was in this period that he published most of his works and gradually established himself in the various intellectual and scientific circles of the Holy Roman Empire. In 1669, he published *Tartaro-Mastix Moraviae per quem rariora et admiranda a natura in faecundo huius regionis gremio effusa, cumprimis tartarus, illiusque effectus morbosi curiose examinantur, et cura tam therapeutica quam prophylactica proponitur* (Vienna, 1669), an extensive work reflecting Hertodt's botan-

4 https://issuu.com/purgly/docs/nobreza_na_hungria/84 accessed 11/09/17. Furthermore, Hertodt's cousin, Matthias Franz Hertodt (in some sources, Franz Matthias Hertodt), was said to be a Hungarian nobleman. http://encyklopedie.brna.cz/home-mmb/?acc=profil_osobnosti&load=11938 accessed 11/09/17. For consistency, we shall call him Matthias Franz Hertodt throughout this book.

5 For detail on this University's medical school, see Zampieri & Zanatta, 2015.

6 Matthias Franz Hertodt was the child of the father's brother (*patruelis*) of Johann Ferdinand Hertodt (Anon., 1676b).

7 Anon., 1676b.

8 In south-western Moravia, in the Czech Republic.

9 Leopold I, 1640–1705, Holy Roman Emperor.

10 Holy Roman Emperor and second son of Leopold I. Lived 1685–1740; reigned from 1711. Charles' elder brother Joseph I (1678–1711) was his predecessor as Holy Roman Emperor.

11 See WorldCat entries for "Hertodt", though there is some confusion over authorship as WorldCat also states Matthias Franz wrote certain editions of *Medicus euporistus*.

12 Entry for Johann Phillip Hertodt von Todtenfeld in the *Encyclopedia of Brno*. (http://encyklopedie.brna.cz/home-mmb/?acc=profil_osobnosti&load=11939, accessed 11/09/17).

ical and geological interests alongside those in medicine. In it, he describes, among other things, rare trees and herbs found in Moravian gardens. The following year he produced another title, the *Opus mirificum sextae diei, id est, Homo physice, anatomice et moraliter in potentiores suas partes dissectus* (Jena, 1670), on the anatomy of the human body, and started working on his third major book, *Crocologia, seu Curiosa Croci, regis vegetabilium enucleatio continens deinque Etymologiam, differentias, tempus quo viret et floret, culturam, usum mechanicum, pharmaceuticum, chymico-medicum, omnibus pene humani corporis partibus destinatum, additis diversis observationibus et quaestionibus crocum concernentibus* (Jena, 1671). *Crocologia* was produced under the patronage, and according to the laws and method, of the *Academia Naturae Curiosorum*, a prestigious scholarly society based in Schweinfurt, which Hertodt had joined on 20th October 1670 using the sobriquet Orpheus.[13] Over the years, several articles by Hertodt appeared in *Miscellanea curiosa medico-physica seu Ephemerides medico-physicae Germanicae curiosae*, a medical journal published by the *Academia* since 1670.

Hertodt's works soon became the object of sharp criticism by one of his colleagues in Brno, Dr Wenzel Maximilian Ardensbach von Ardensdorf (1630–1684), who was in competition with Hertodt for the position of municipal physician of Brno.[14] In 1671, shortly after the publication of *Crocologia*, in an attempt to show Hertodt's ineptitude and lack of originality, Ardensbach published *Tartaro-Clypaeus, excipiens Tartaro-Mastigem Hertodtianum, cui accredit discursus in Opus mirificium VI. diei, item Animadversatio in Crocologia Hertodtianam* (Prague, 1671), where he castigated Hertodt's works (see Appendix 2 for our translation of Ardenbach's tract about *Crocologia*). Hertodt responded with *In methamorphosim ridiculam Ardensbachianam hominis pulicis* (Jena, 1671). Despite Ardensbach's attempts to diminish him, Hertodt's prestige continued increasing and in 1686, he was appointed personal physician of the Emperor Leopold I, just like his cousin and mentor Matthias Franz Hertodt (whom Hertodt mentions with respect and fondness several times in *Crocologia*). Our author's new position required him to move to Vienna (almost 90 km south of Mikulov) and gained him more favour at court, so that in February

13 The *Academia Naturae Curiosorum* ("an academy of curious inquirers into nature" according to the Royal Society, Hutton et al., 1809, p. 561) was founded in Schweinfurt on 1st January 1652 and was later given the name *Leopoldina* by Emperor Leopold I. The *Academia Leopoldina* is still active and now called the German National Academy of Sciences or simply the Leopoldina.

14 Sajner, 1974, p. 1139.

1701 he was knighted and given the title "von Todtenfeld".[15] He died in 1714. No known portraits of him survive.

1.2 *Description of* Crocologia

Crocologia is a rare book, only held by around 40 public repositories worldwide, mostly at academic institutions and state libraries.[16] To write it, Hertodt drew together a vast corpus of formulae, advice and observations from published sources, personal communications from other physicians, and his own findings. In the book's original 283 pages, Hertodt cited over 200 authors, ranging through western medical history from Hippocrates, Galen and Dioscorides, right up to his contemporaries like Rolfinck, Zwelfer, Mynsicht, Lauremberg and Duchesne. In addition, the *Pharmacopoeia Augustana*, the official standard reference book for Vienna and the Austrian provinces from 1618, is the source of some of *Crocologia*'s formulae.[17] As well as saffron, Hertodt called for several hundred other ingredients: herbal, mineral, animal and human, sourced from land, sea and rivers from the Arctic to the tropics, over all five continents then known; Hertodt also mentions many man-made chemicals (see Glossary for more details).

Four typefaces occur in the book: Roman, Italic, Greek and Fraktur; numerous alchemical symbols are also used. Hertodt used Fraktur to quote German terms and for labelling medicines. Although not always consistent, he emphasised the word saffron, and its derivatives, throughout the book by using italics.

Crocologia contains an engraved title plate, showing the legend of Crocus and Smilax, and also an engraved, folded plate with eight depictions of crocuses. These include two views of saffron, as well as other species, showing differences in flower, leaf and corm morphology. The saffron images look broadly accurate at first glance, except that the corm tunics are wrongly depicted as being pubescent, and one of the saffron flowers has seven tepals instead of the normal six.

Beginning with a dedication to Leopold I, *Crocologia* next presents a cabbala by Matthias Franz Hertodt, in which certain letters are capitalised to give the date 1670 (the year of *Crocologia*'s writing), and letters in the two further lines are given numeric scores which added together in particular arrangements, again make 1670. There are then citations by three of the author's contemporaries, two of whom were his co-fellows of the *Academia*.

15 His cousin Matthias Franz Hertodt was also knighted two years earlier (February 1699) (Kneschke, 1863, vol. 4, p. 337).

16 Data from WorldCat. The country with the most libraries holding *Crocologia* is Germany.

17 Sonnedecker, 1963, p. 88.

Each chapter, and several sections, are introduced with stories from classical literature, folklore or the bible. *Crocologia* starts with a chapter on the origin of saffron's name, covering the legend of Smilax and Crocus as related by Ovid. Hertodt also gives us saffron's botanical description and its names in different languages. In chapter II, related species are discussed, both those which flower in the spring and the autumn. Hertodt draws heavily on the works of botanists including Clusius, Dodoens, Tabernaemontanus, Lobel and Bauhin, as well as quoting from the wonderful *Hortus Eystettensis*.[18] We consulted the UK's National Collection Holder for the genus *Crocus*, who was able to assign modern botanical names to some of Hertodt's listed plants. Interestingly, Hertodt also includes safflower in this section, despite its being botanically unrelated to true saffron. This is in stark contrast to his colleague's (Georg Christoph Petri) published view that safflower is a kind of thistle, which is much closer to the botanical truth. Confusion between saffron and safflower is still commonplace today (see below).

In chapter III, brief detail is given on saffron's native place and flowering time, in which Hertodt mentions his own saffron production in 1669. Chapter IV covers saffron's cultivation methods, and Hertodt had plenty of choice in his research material, for in the fifteenth and sixteenth centuries "the cultivation of saffron was so important an article in the European husbandry, that it was omitted by no writer on that subject."[19] There is much emphasis on the control of moles and mice, including descriptions of traps, which Hertodt's rival and critic, Ardensbach thought "unseemly" for a physician to write. The age-old problem of choosing good saffron, and identifying adulterants is the subject of chapter V. We detail this from a modern perspective in section 2.7.

Next, Hertodt moves away from botany and into pharmacy. Chapter VI is devoted to the properties of saffron, both good and bad. Quoting from several sources, our author states that saffron is a warm and dry drug, but to which degree it possesses these properties, there is no consensus. Chapter VII concerns the many different preparations of saffron, some of which remained in use into the twentieth century. Most of Hertodt's formulae require saffron threads, the usual part of the plant in commercial use. Significantly, however,

18 The *Hortus Eystettensis* was a sumptuously illustrated catalogue of the plants in the Prince Bishop's famous garden at Eichstätt, Bavaria. It was printed on the largest paper then available, and released in two versions. The black-and-white edition contained text as a reference for apothecaries etc., and a hand coloured version was without text. The *Hortus Eystettensis*' author, Basil Besler, was able to buy himself a large new house from his profits, and the house's price equated to just five coloured copies of the book. (British Library https://www.bl.uk/collection-items/hortus-eystettensis accessed 25/09/2019).

19 Johnston, 1797, p. 286.

some preparations call for saffron flowers (after the threads have been removed, e.g. a water distilled from the flowers, p. 69) and saffron corms (e.g. liniment of saffron, p. 83). It is interesting to speculate whether there was a formal trade in flowers and corms for medicinal use, or whether they were employed only in the saffron-growing areas during the season they were available fresh.[20] Today, saffron flowers and corms are beginning to attract attention as novel targets for biochemical and therapeutic research, see section 4.

Hertodt afterwards takes us on a head-to-toe journey of the human body over fourteen more chapters, covering diseases and therapeutic formulae containing saffron. Chapters VII to XXII (there is no chapter XI) include diseases and conditions of the brain, eyes, ears, chest, heart, stomach, intestines, liver, spleen, bladder and kidneys (including impotence), female reproductive system, and "external parts" (a mixed bag ranging from erysipelas to gunshot wounds). There is a chapter on fevers, the greater part of which concerns the plague. Hertodt includes the formula for a prophylactic tincture which he had been prescribed by his cousin whilst living in Mikulov during the 1665 outbreak of plague there.

Given Hertodt's standing and typical patients, plus the expense of exotic ingredients and the complex, lengthy methods required for their preparation,[21] *Crocologia* is clearly a treatment manual for the well-to-do. Hertodt offers only two recipes for the less affluent: his ointment of saffron, which has just four ingredients and is good for the "neglected sick" and a simple infusion of hyssop with saffron to treat poor patients with asthma.

Lastly, Hertodt discusses saffron's "mechanical uses" and proposes answers to four saffron-related questions current at the time of *Crocologia*'s publication. One of these questions was whether or not saffron can colour a foetus *in utero*. The results of Hertodt's investigations into this were first published by him in the *Miscellanea curiosa medico-physica* in 1670,[22] and have been cited as the first evidence of a link between the maternal circulation and the foetus.[23]

Crocologia finishes with a five-page index. There is no reference list in the original book, but we have constructed one (see Appendix 1).

---

20 SF is not aware of saffron tepals or corms being used medicinally in the UK. Saffron flowers and tepals can be dried and thus preserved, but this must be done on the day of picking because otherwise they quickly turn into a revolting mush. Corms can last for several months before they completely shrivel (SF pers. ob.).

21 For example, see Tomlinson, 1657.

22 Hertodt, 1670, p. 157.

23 Needham, 1959, p. 180. Needham incorrectly states that Hertodt injected saffron into the maternal circulation, but Needham was not able to access a copy of *Crocologia* to check his statement.

2 Saffron

2.1 *What Saffron Is, and Is Not*

Saffron, the focus of *Crocologia*, is surrounded by many misconceptions and misinformation. Saffron is the dried stigmata and style branches (together called the "threads") of the saffron crocus, *Crocus sativus* L. (Family: Iridaceae).[24] Historically, saffron was given different officinal designations e.g. *crocus Anglicus*, *crocus Hispanicus*, *crocus Gallicus* etc.[25] Potentially misinterpreted as Linnaean binomial names, they are officinal names that instead relate to the saffron's origin. In *Crocologia*, when Hertodt does specify provenance, he calls for *crocus Austriacus* ["Austrian saffron" (produced in Lower Austria[26])] or *crocus orientalis* ["eastern saffron"].[27] Three other plants confused with saffron are:[28]

- Safflower, *Carthamus tinctorius* L., (Compositae), or bastard saffron, American saffron, dyer's saffron,[29] false saffron, saffron thistle,[30] Canaries saffron,[31] Mexican saffron, and in the older literature: *crocus crispus*, *crocus fatuus*, *crocus hortensis*, *crocus hortulanus*, *crocus saracenicus*, *crocus sylvestris*,[32] *crocus occidentalis*,[33] *carthamus officinarum flore croceo*[34] and Spanish

24 Sometimes saffron is referred to as "filaments", but this word has a precise definition in botany that does not apply to style branches and stigmata. "Patti" saffron is the dried stamens of the saffron plant, sold in India (Serrano-Díaz et al., 2013).

25 Pereira, 1840, footnote p. 674. Certain chemical preparations in *Crocologia* also contain the word "crocus" e.g. crocus metallorum (see Glossary).

26 Dr Bernhard Kaar, Wachauer Safran, pers. comm., 14/10/18.

27 Not Oriental saffron; see p. 30 for our reasoning behind this translation. The characteristics of domestically-produced and imported saffrons are discussed from an Austrian perspective by Alb, 1854, p. 89 et seq.

28 A further saffron-substitute plant briefly imported into England from the Cape of Good Hope in Victorian times was "Cape saffron", described as toad-flax-like flowers that gave a good saffron colour in aqueous infusions and also had a saffron-like aroma (Wood & Sharp, 1866, p. 4). This is *Sutera atropurpurea* (Banks) Hiern (syns. *Lyperia crocea* Ecklon and *L. atropurpurea* Benth.) (Scrophulariaceae), also known as African saffron (Seidermann, 2005, p. 354). An additional plant used like true saffron is saffron-of-the-Andes or saffron root, *Escobedia grandiflora* (L.f.) Kuntze (Orobanchaceae; syn. *E. scabrifolia*), whose deep orange root is used as a spice in Peru, Colombia and Venezuela (Seidermann, 2005, p. 148).

29 Francis, 2011, p. 12.

30 Seidermann, 2005, p. 86.

31 SF, pers. ob.

32 Ernstingium, 1770, p. 350 et seq. Additionally, this source lists *crocus vasicus* and *crocus vasius* as names for the dye-plant madder (*Rubia tinctorum* L. (Rubiaceae)).

33 Zedler, 1793, p. 1680 et seq.

34 Miller, 1759, see under Carthamus.

saffron.[35] Safflower is a thistle-like plant whose yellow or orange florets are a dyestuff, or a poor saffron substitute often fraudulently sold as fake saffron (see section 2.7). Its seeds can be pressed for oil.

- Turmeric, *Curcuma longa* L., (Zingiberacae), or Indian saffron, Bourbon saffron,[36] saffron root, *Safranwurzel* in German[37] and in the older literature, *crocus indicus*.[38] The fleshy, yellow rhizome, usually sold in dried and powdered form, is rich in curcumin. Both turmeric and purified curcumin (the yellow food colouring E100) are used for culinary, medicinal and cosmetic purposes.[39]
- Meadow saffron, *Colchicum autumnale* L. (Colchicaceae), or autumn crocus (a misnomer: it is in a different family from crocuses) and naked ladies. A toxic autumn-flowering plant with crocus-like flowers. Older writers make a clear distinction between *Colchicum* and saffron, both of which were used in medicine. However, superficial similarities between mauve forms of *Colchicum autumnale* and saffron, and the use of the term "meadow saffron" means today there is often confusion between the two.[40]

2.2 *Uses of Saffron*

Saffron's major use today is in cookery, where it imparts a rich golden-yellow colour, a honey-like flavour and a unique aroma to sweet and savoury foods; it is the signature ingredient of several dishes e.g. risotto alla Milanese (Italy), paella (Spain), Cornish saffron cake (UK), and lussekatter (Sweden), and is widely used throughout the cuisines of India and Iran.

Saffron is also employed in medicine in many different traditions over the world. Hertodt amply summarises saffron's virtues according to classical writers, see ch. VI, and its use continued in orthodox western medicine quite some time after *Crocologia*'s publication.

In modern European herbal medicine, saffron is a sedative, antispasm and anti-asthma agent, according to the German Commission E monograph of

35 Parkinson, 1629, p. 329.
36 Thiercelin, 2003, p. 151.
37 Seidermann, 2005, p. 126.
38 Ernstingium, 1770, p. 350.
39 Dhakal et al., 2016.
40 On many occasions, SF has heard members of the public state they have saffron growing in their gardens when on further investigation, they actually have *Colchicum autumnale*, which is more readily available than saffron plants, and, unlike saffron, can naturalise under UK conditions. Worryingly, people have even gathered "saffron" from *Colchicum* based on this dangerous assumption, and been made ill after cooking with it.

1987.[41] However, although many of the Commission E monographs have been updated, neither the ESCOP (European Scientific Cooperative on Phytotherapy) nor HMPC (the EU's Committee on Herbal Medicinal Products) have produced new monographs on saffron as yet.[42] Saffron also features in homeopathy. It was not amongst the first "provings" of substances, but was later studied by Stapf in the early part of the nineteenth century and its various effects and uses in homeopathy laid out.[43] Saffron is still used in homeopathy today.[44]

Asian medicine includes the use of saffron. In the Ayurvedi, Unani and Sidha systems, saffron is regarded as an: anodyne, antispasmodic, carminative, aphrodisiac, diaphoretic, emmenagogue, expectorant and sedative; and is used against: fever, smallpox, colds, headache, insomnia, depression, asthma and tumours.[45] Traditional Chinese medicine employs saffron "to stimulate blood flow and relieve pain by removing stagnated blood."[46]

Outside formal systems of medicine, saffron was and still is an important home remedy. In the UK, for example, saffron was a common domestic cure especially for measles. In 1945 (and later), saffron tea was still being used for this purpose in the Eastern Counties of England.[47] Saffron also had a darker side though, sometimes being given in deliberately high doses as an abortifacient.[48] This could prove fatal to the woman concerned.[49] Immigrants have brought

41 Blumenthal et al. (1998) *The Complete German Commission E monographs: therapeutic guide to herbal medicines.* American Botanical Council. American Botanical Council's website http://cms.herbalgram.org/commissione/Monographs/Monograph0322.html accessed 25/09/2019.

42 Dr Christopher Etheridge, Chair of the British Herbal Medicine Association, pers. comm., 20/09/2019.

43 Hempel, 1846, pp. 166–181.

44 Francis Treuherz, Homeopath, pers. comm., 17/09/2019.

45 Quattrochi, 2016, p. 1176.

46 Tong et al., 2015.

47 Newman, 1945, p. 355. Saffron was also an ingredient of cakes in the counties of Essex and Cornwall at this time. Saffron tea was a common domestic remedy for measles used within living memory, probably into the 1950s, in Eastern England. SF has been told several times of how, even if the intended purpose was for cookery, saffron was always bought from a pharmacist rather than a grocer. The traditional practise of using saffron to bring out the rash of measles was dismissed in 1932 as a "curious old delusion" (Harrison Martindale, 1932, p. 850).

48 For example, an English regional newspaper in 1834 reported on the case of John Coe, convicted of "administering saffron to one Henrietta Freeman, for the purpose of procuring a miscarriage" for which he was sentenced to 7 years' transportation. *Norfolk Chronicle,* 22nd March 1834, p. 4.

49 The symptoms of saffron poisoning after administration as an abortifacient are described by Anon., 1933.

their own traditional uses to new areas e.g. the Bangladeshi community of West Yorkshire, UK, uses saffron (*zafraan*) as a carminative, and also "to ward off evil and help people afraid of the dark … [saffron is] added to a small amount of oil and rubbed on the skin before going to bed."[50]

Saffron also has minor usage as a dyestuff and in perfumery.

2.3 *Botany & Origin*

Crocus sativus is a geophyte. It grows from an underground corm of up to 50 mm in diameter covered by a fibrous tunic extending upwards from the corm apex for 50 mm or more. The plant has 5–11 leaves, normally only 1.5–2.5 mm wide but reaching up to as much as 500 mm in length. On average, 1–4 flowers develop from each corm,[51] each with six tepals up to 50 mm long, lilac to mauve in colour, with darker violet veins and sometimes a violet stain in the throat. These flowers are fragrant (not a universal characteristic of crocuses), with a honey-violet scent, highly attractive to insect pollinators, and, unusually for a crocus, they stay open at night.[52] Inside the flowers are three stamens (*Colchicum* has six) and a single style divided into three deep scarlet clavate branches, which is the part harvested as saffron threads.[53] Saffron's corms and its style branches are the largest amongst the crocuses.

Compared with other members of the *Crocus* genus, saffron is unusual.[54] It is a sterile triploid (2n = 3x = 24) hybrid plant,[55] unknown in the wild.[56] Its

50 Pieroni et al., 2010, p. 135.

51 A single corm can produce up to 13 flowers (SF, pers. ob.).

52 The scent of fresh saffron flowers is different from that of dried saffron spice (SF, pers. ob.) and has a different volatile composition to dried saffron (Lim, 2014). Bees are strongly attracted to saffron flowers, and are often found in a soporific state inside them (Basker, 1999, p. 47; SF, pers. ob.).

53 Saffron flowers with supernumerary organs exist, e.g. with 4 style branches, 4 stamens and 8 tepals. In France, one M. Chappellier of Paris attempted to propagate a monstrose form of saffron crocus where the tepals were replaced by extra pistils and stamens (Maw, 1886, p. 19 & plate D). Despite the obvious advantages of such flowers, they are sporadic mutations only, and cannot be selected for by propagating from the corms that produced them. In contrast, an ornamental, *C. chrysanthus* cv. 'Goldmine', is a form with stable vegetative inheritance of double flowers (Rukšāns, 2010, plate 125).

54 The genus *Crocus* was until comparatively recently considered to have around 88 species, but new botanical field work and molecular genetic studies have revealed many novel species, and raised old subspecies to species rank. It has been estimated that with the continuation of such work, the final species count may reach 300–400 (Rukšāns, 2017, p. 26, quoting a communication from H. Kerndorff).

55 There have been reports of capsules and seeds being produced very rarely (Mathew, 1999).

56 Saffron is unable to maintain itself indefinitely without human intervention in the form of corm lifting, division, and replanting. Under UK conditions, saffron plants did not persist

sterility presents two challenges to growers: (i) saffron can only be propagated by lifting and dividing the corms (a human intervention), and such vegetative reproduction is slower than sexual reproduction via seeds would be, (ii) there have been no opportunities to breed better cultivars by crossing different lines to make new hybrids, as has been successfully done with other crops to improve traits like yield, quality, disease resistance, salt tolerance etc.[57]

Saffron threads do not survive in the archaeological record; however, saffron is depicted on Minoan pottery and frescoes dating to around 1600 BC and linguistic evidence suggests that saffron was brought into cultivation long before the classical period.[58] Although the Zagros mountains in Iran[59] have been put forward, the Mediterranean area is saffron's correct birthplace.[60]

In his classification of the genus, Mathew placed *C. sativus* in Crocus series Crocus (the "saffron group") along with several other species, based on morphological characters.[61] The most similar of these plants to saffron is *C. cartwrightianus* Herbert, a fully-fertile plant of open rocky hillsides from sea level to 1,000 m in Greece. Its style branches are 10–27 mm long in contrast to saffron's at 22–35 mm, and it is a more variable species, including white-flowered forms.[62] *C. cartwrightianus* was proposed to be the original source of "saffron",[63] from which *C. sativus* was selected for its larger threads, and domesticated in Bronze-Age Crete.[64]

---

beyond three years in an experimental abandoned plot (SF, pers. ob.), though it is likely the plants would survive for longer in areas with a drier climate.

57 Plant breeding has made a gigantic contribution to yield increases in most common agricultural crops worldwide. For example in wheat in the UK, yields per hectare have increased threefold since 1948. Between 1948 and 1982, 57% of that increase was due to breeding new cultivars; between 1982 and 2007, 93% of the increase was due to plant breeding (*Plant Breeding Matters*, Spring 2012, p. 1). The rest of the increase is accounted for by factors including better disease control, crop nutrition etc. With saffron, there has been some success by propagating from selected corms (Agayev et al., 2009), but still no true, named, cultivars exist.

58 Zohary & Hopf, 2000, p. 207.

59 Ghorbani & Koocheki, 2017.

60 Vavilov, 1992, p. 341; Mathew, 1999.

61 Mathew, 1999.

62 Mathew, 1999. *Crocus cartwrightianus* is an easier species to grow under UK conditions than *C. sativus* and the former was given an AGM (Award of Garden Merit) by the Royal Horticultural Society, meaning it is a reliable performer in British gardens.

63 *Crocus cartwrightianus* threads are still gathered today as "saffron" (Seidermann, 2005, p. 122; Rukšāns, 2010, p. 99). Several additional *Crocus* spp. are used as local substitutes for saffron, some also as saffron adulterants (SF, in prep.).

64 Negbi, 1999, p. 1.

There is minimal genetic diversity in saffron accessions from different countries, meaning that the same clone of saffron is grown around the world and probably this clone has only arisen once.[65] This is remarkable amongst crop plants. It was recently postulated that saffron was an allotriploid hybrid derived from *C. cartwrightianus* and *C. pallassii* subsp. *pallassii* or close relatives.[66] Newer work has reached a different conclusion: that saffron is an autotriploid derived solely from *C. cartwrightianus*.[67] This finding raises the exciting possibility that saffron can be "re-synthesised" by crossing *C. cartwrightianus* accessions, and that the world's first saffron breeding programmes may be set up in the future.

2.4 *Production Areas*

Spread by invaders, occupiers, immigrants, and sometimes single entrepreneurs, saffron is now grown in all the world's continents except Antarctica. But the precise details of its introduction (and sometimes decline or disappearance, and occasionally reintroduction) into different countries is complex and not well studied.[68]

Spain was formerly the world's largest saffron producer, in 1930 growing 11,000ha of the plant, yielding over 80t of the spice.[69] Various factors including rural depopulation, high labour costs and lack of mechanisation caused plantings to be abandoned and fears were raised that Spanish-grown saffron would eventually disappear. With EU PDO (Protected Denomination of Origin) status now granted, genuine Azafrán de la Mancha (grown in Albacete, Cuidad Real, Cuenca & Toledo; on 132ha, 2019 figures[70]) is undergoing a welcome revival.

In 2004, total European saffron production stood at approximately 6.8t, or just 4% of global production.[71] As well as Spain, European saffron producers include Greece (4.5t p/a; Europe's largest grower),[72] Italy,[73] Switzerland (in

65 Based on studies using AFLP (amplified fragment length polymorphism) markers (Busconi et al., 2015).
66 With IRAP (inter-retroelement amplified polymorphism) markers (Alsayied et al., 2015).
67 Nemati et al., 2018 & 2019; Schmidt et al., 2019.
68 Negbi, 1999.
69 Alonso Díaz-Marta et al., 2006, p. 22.
70 *Denominación de Origen Protegida Azafrán de la Mancha, Spain.* http://doazafrandelamancha.com/images/documentos/Estadisticas/Calificado_Provincial_2019.pdf accessed 2/3/2020.
71 Alonso Díaz-Marta et al., 2006, p. 21.
72 Goliaris, 1999; Christodoulou et al., 2015.
73 Tammaro, 1999.

Mund),[74] Austria (the Wachau region),[75] England,[76] and France (several areas including a traditional centre in the Gâtinais).[77] Though it does not produce the spice for sale, the Netherlands is now a large grower and exporter of saffron corms.[78]

Global trade in saffron, worth an estimated $238 m in 2017, is today dominated by Iran.[79] Most of the crop comes from the north-eastern province of Khorasan.[80] Just over half of Iran's saffron production is bought by Spain, the world's largest saffron importer and the second largest exporter,[81] mostly for repackaging into small retail jars, tins and sachets. Iranian saffron plantings, already at over 87,000ha in 2014,[82] have undergone colossal expansion and the *Tehran Times* claimed that 400t was produced in the year beginning 21st March 2018, against 236t (worth $326 m) the previous year.[83]

Some other current saffron-production areas are:

- Turkey, around Safranbolu; now thriving but the crop had previously shrunk to just one grower on 0.1ha.[84]
- Azerbaijan, on the Absheron Peninsula near Baku.[85]
- Afghanistan; recently re-introduced as an alternative to opium poppies.
- Kashmir; but production is in decline.[86]
- China; grown in 20 provinces since the 1970s, 1t produced in 2013, mostly in Zhejiang Province.[87]

---

74 *Mund, Switzerland, saffron website.* https://www.aop-igp.ch/en/munder-safran-aop/ accessed 20/9/2019.

75 *Wachau, Austria, saffron website.* https://www.safranmanufaktur.com/wachauer-safran manufaktur/ accessed 20/9/2019.

76 Including on SF's small farm. For a history of saffron in England, see Francis, 2011.

77 Thiercelin, 2003.

78 *North American Center for Saffron Research & Development, University of Vermont. Presentation by Dutch corm growers, Roco Saffron.* https://www.uvm.edu/~saffron/Resources/Pres entations/SaffronProduction-Hans.pdf accessed 20/9/19.

79 *Economic Complexity Observatory.* https://oec.world/en/profile/hs92/091020/#Exporters accessed 12/09/2019. (Simoes, A.J.G. & Hidalgo, C.A. (2011) *The Economic Complexity Observatory: An analytical tool for understanding the dynamics of economic development.* Workshops at the Twenty-fifth AAAI Conference on Artificial Intelligence).

80 Kafi et al., 2006.

81 https://oec.world/en/profile/hs92/091020/#Exporters accessed 12/09/2019.

82 Gorbani & Koocheki, 2017.

83 *Tehran Times.* https://www.tehrantimes.com/news/432135/Saffron-output-increases-15 accessed 21/01/19.

84 Negbi, 1999.

85 Azizbekova & Milyaeva, 1999.

86 Husiani et al., 2013.

87 Tong et al., 2015a; 2015b.

- Morocco, around Taliouine.[88]
- North America: arrived in the USA before 1730 with religious groups like the Amish, Mennonites and Schwenkfelders, still grown in Pennsylvania (especially Lancaster County) by their descendants.[89] Boutique crops in Canada.[90]
- South America, e.g. Chile.[91]
- Oceania: New Zealand and Australia.

2.5 *Cultivation and Drying*

Saffron production methods vary slightly across the globe based on weather patterns and local traditions. The plant is a perennial, grown as an annual (e.g. in Navelli, Italy, where the corms are lifted and replanted every year)[92] or pluri-annual crop (where the saffron plants are in place for 3–10 or more years before lifting and replanting). Sometimes the plant is grown alone, but it may also be under-planted in orchards[93] and has been grown experimentally amongst solar panel arrays.[94]

The following description and timings are for crops in the northern hemisphere.

Saffron corms are planted during the plants' summer dormancy season, ranging from May to mid September.[95] The beds are weeded in readiness for the emerging flowers in autumn and may also be irrigated. Flowering is triggered by a drop in air temperature[96] and commences from late September[97] to early November.[98]

Harvesting is the most labour intensive period of the year, lasting from 10 days (Spain)[99] up to 4–6 weeks (Norfolk, UK).[100] The flowers are gathered by

---

88 Ait-Oubahou & El-Otmani, 1999.

89 Williard, 1998, p. 132 et seq.

90 *CBC News*. https://www.cbc.ca/news/canada/montreal/canada-s-first-saffron-farm-expecting-excellent-crop-1.2776626 accessed 29/01/19.

91 *North American Center for Saffron Research & Development, University of Vermont. Presentation by Dutch corm growers, Roco Saffron*. https://www.uvm.edu/~saffron/Resources/Presentations/SaffronProduction-Hans.pdf accessed 20/9/19.

92 Tammaro, 1999.

93 Ait-Oubahou & El-Otmani, 1999, p. 90.

94 *Solar Power World Online*. https://www.solarpowerworldonline.com/2019/02/solarsaffron-experiment-could-pave-the-path-to-more-dual-use-farms/ accessed 15/02/2019.

95 Alonso Díaz-Marta et al., 2006, p. 25.

96 Molina et al., 2004. Traditional French folklore states that saffron begins flowering there "at the third rain of autumn after August 15th" (Thiercelin, 2003, p. 67).

97 In SF's crop, one of the most northerly in Europe.

98 Alonso Díaz-Marta et al., 2006, p. 28.

99 Alonso Díaz-Marta et al., 2006, p. 27.

100 SF, pers. obs.

hand, typically early in the morning, sometimes before they open, then taken indoors for the threads to be removed, again by hand. One factor influencing the saffron's strength is how the threads are removed, in other words, how much of the lower yellow or colourless style tissue is left attached to the red style branches. The "wet saffron" must be dried straight after harvesting each day both to preserve the spice and to develop its characteristic organoleptic profile (see below, under Biochemistry). Gathering, thread removal and drying is repeated daily during harvest. It is an exacting and time-consuming process with an unpredictable workload, which is why saffron is often grown on small farms using family labour (sometimes organised into cooperatives).

The empty flowers are a waste product, once discarded in England[101] though traditionally fed to dairy cows in France,[102] and have been the focus of research into novel medicinal uses (see below), or employed as a dyestuff.[103]

The saffron threads may be sun- or air-dried outside, or desiccated indoors using artificial heat,[104] e.g. in mainland Europe by suspending the saffron in silk or wire mesh sieves over a charcoal brazier, or more recently, a butane burner.[105] Here, the saffron can reach temperatures up to 100°C.[106] A variant method in Sardinia involves wetting the saffron with extra virgin olive oil (the *feidatura* process), before drying it on planks of wood placed at a fireside.[107] In Greece the threads are sprinkled on large silk sieves on racks in a dedicated drying room at 25–30°C for 12–24h. Today, many producers use electric food dehydrators, but some freeze-dried saffron is also produced.[108] Saffron's quality is strongly affected by its drying, and subsequent storage, conditions.[109] Approximately 150–200 flowers must be processed to yield 1g of dried saffron.[110] This also generates 63g of bio-residues (tepals, stamens and styles) per gram of saffron.[111]

101 A decree of 1575 from the Essex town of Saffron Walden (UK) forbade growers from throwing saffron flowers and other rubbish in the river there (Holmes, 1988, p. 24). Now, the tepals from SF's crops are used in novel skin-care products.

102 Thiercelin, 2003, p. 71.

103 Raja et al., 2012. On pashmina shawls, saffron-flower-waste dye gave a greenish yellow colour and conferred antimicrobial properties.

104 Carmona et al., 2005.

105 Alonso Díaz-Marta et al., 2006, p. 39.

106 Thiercelin, 2003, p. 116.

107 Alonso Díaz-Marta et al., 2006, p. 37.

108 SF, pers. obs.

109 Carmona et al., 2005.

110 Pereira 1840, p. 674, calculated that 1 grain of saffron (approximately 65mg) contained the stigmata and styles of nine saffron flowers.

111 Serrano-Díaz et al., 2013.

The above processing techniques produce "hay saffron" (*crocus in foeno*[112]), where the threads are separate from each other. This may then be further processed into powdered saffron. In England, a alternative product was also once made: "cake saffron" (*crocus in placentá*[113]), a compressed sheet of intermingled saffron threads created in special kilns by the application of pressure and heat over a 24-hour period.[114] Cake saffron production is now extinct and no known saffron kilns remain, but details of their size, construction and usage in England are recorded.[115] Persian, Russian and Baku saffron was once sold in cake form too.[116] Another kind of compressed saffron was produced in seventeenth century Germany, where the spice was made into little balls, "by mingling it with a little Honey; which thoroughly dried, they reduce to Powder, and sprinkle it over their Sallets for a noble Cordial."[117]

After harvest has finished, the plants continue growing throughout the winter. By the commencement of the dormancy period in around May, each mother corm will have developed a number of daughter corms and died itself. The dead saffron leaves, and weeds of old saffron plots, are cut for hay in some areas.[118] At lifting time, unwanted corms may be fed to livestock,[119] but in France were also formerly used as a source of starch in chocolate manufacture.[120] The corms have also "been proposed to be made into bread in times of scarcity."[121]

Saffron has only relatively recently attracted the kind of agricultural research that has been applied, for a much longer period and with great success, to broad-acre crops e.g. wheat. There is now a much better understanding of saffron's fertiliser requirements, and of the crop's pests and diseases, for example. In Spain, machinery is now being developed to mechanically harvest the crop, and some growers are reported to produce saffron in glasshouses under controlled environment conditions.[122]

112 Pereira, 1840, p. 674.
113 Pereira, 1840, p. 674.
114 For clarity, "cake saffron" is a dried, compressed mass of saffron threads, whereas "saffron cake" is a cake or bread flavoured with saffron.
115 Bradley, 1727, p. 151 et seq.; Bradley, 1736, p. 158 et seq.
116 Abl, 1854, pp. 91–93.
117 Evelyn, 1699, p. 68.
118 In Austria, these withered saffron leaves were called *Soper*, and were believed to be a good livestock feed for encouraging lactation (Abl, 1854, p. 89).
119 Ait-Oubahou & El-Otmani, 1999, p. 89; Tammaro, 1999, p. 56; Thiercelin, 2003, p. 71.
120 Thiercelin, 2003, p. 71. The same author, p. 198, quotes a work from 1766 stating that saffron corms have a sweet taste.
121 Redwood, 1847, p. 543.
122 Moratella-López et al., 2019b.

2.6 *Saffron Biochemistry*

Saffron's biochemistry is a large and complex subject, presented here distilled into its very simplest form.[123] Saffron threads have been the focus of most research on the plant. The threads' three major bioactive compounds are:

- The carotenoid pigments crocins, the most common of which is crocin 1 (α-crocin), an ester with the sugar, gentiobiose. Crocins' precursor is crocetin.[124] Normally carotenoids are lipophilic, but crocins are uniquely hydrophilic carotenoids.[125] They are powerful colorants and their presence can be detected by eye in aqueous solutions at 1 part per million.[126] Crocins are stable at high temperatures, giving them much versatility in cookery.[127] Crocins account for 6–16% of saffron's dry matter. They have strong antioxidant properties, quenching free radicals and protecting cells and tissues from oxidative damage.[128] When oxidation happens (which can be catalysed by exposure to light), crocin's colour fades.[129]
- Picrocrocin, a product of zeaxanthin degradation, largely responsible for saffron's bitterness and flavour. Picrocrocin represents 1–13% of saffron's dry matter content.[130] It is thought picrocrocin helps protect the stigmata from insect browsing.[131] Picrocrocin has so far only been indentified in *Crocus* species.[132]
- Safranal, saffron's "volatile oil", the result of picrocrocin de-glycosylation. Safranal is largely responsible for saffron's characteristic hay-like aroma and its production is stimulated by the proper drying of saffron, mediated by thermal breakdown and/or enzymatic degradation.[133]

Saffron tepals contain, amongst many other substances, the flavonols quercetin, kaempferol, galangin and anthocyanins e.g. delphinidin.[134] Kaempferol, an antioxidant with better pharmacokinetic properties than the popular food supplement quercetin, is present in very high levels in tepals, making them a

123 For more information, see e.g. Bagur et al., 2018.
124 Melnyk et al., 2010. Crocin is hydrolysed into crocetin in the intestinal tract after eating saffron (Bagur et al., 2018) or oral administration of crocin (Moratella-López et al., 2019b).
125 Finley & Gao, 2017.
126 McGee, 2004, p. 423.
127 Edwards, 2000, p. 71.
128 Melnyk et al., 2010.
129 Basker, 1999, p. 47. Crocin was formerly called polychroite, because of its various colour changes after the addition of different reagents (Murray, 1824, p. 105).
130 Melnyk et al., 2010.
131 McGee, 2004, p. 423.
132 Bagur et al., 2018.
133 Basker, 1999, p. 47.
134 Lim, 2014, p. 84; Moratella-López et al., 2019b.

hitherto overlooked valuable resource.[135] Indeed, saffron tepals contain more than 100 times the kaempferol levels of foods considered "rich" in this substance.[136] The tepals are also high in iron.[137] Saffron corms and cell suspension cultures have also been the target of studies.[138]

2.7 *Substitutes, Adulteration and Fraud*

Saffron's high price has always exerted pressure for a cheaper alternative product. If substitutes (safflower or turmeric) are honestly labelled, then consumers can make an informed choice whether to buy or not. Problems arise when entirely fake saffron is sold as genuine, or when real saffron has been admixed with other substances to make it go further or appear better. Additionally, weak or ungraded saffron may be labelled as being of the highest quality, or the saffron claimed or implied to have a certain provenance despite its being grown elsewhere. Pliny famously wrote, "There is nothing adulterated so much as saffron".[139] This practice spawned the world's first food standards legislation, the 1358 Nuremburg *Safranschou*, with capital punishments for offenders.[140]

Adulteration occurred in Hertodt's time and continues today. A common example of counterfeiting is where safflower florets or ground turmeric[141] are passed off as real saffron to unsuspecting tourists in markets.[142] Worldwide, other "bulking agents" and adulterants recorded in saffron include:[143]

- Botanicals: e.g. maize silks, saffron stamens and other floral waste, marigold (*Calendula officinalis* L.), dyed willow (*Salix* sp.) or grass roots, *Arnica montana* L., annatto (*Bixa orellana* L.), *Crocus vernus* (L.) Hill, *Hemerocallis* sp., *Gardenia jasminoides* J. Ellis,[144] *Nelumbo nucifera* Gaertn., *Chrysanthemum x morifolium* Ramat.

---

135 Zeka et al., 2015.

136 Moratella-López et al., 2019b. It is important to note that, as far as the authors are aware, the tepals or whole flowers are not authorised for food use.

137 Serrano-Díaz et al., 2013.

138 See Lim, 2014.

139 Pliny, *Natural history*, book XXI, ch. 17.

140 Williard, 2001, p. 102. In France in 1550, Henry II issued an order to prevent adulteration of saffron. He recognised not only the human health implications of adulterated saffron, but also its effects on the reputation of French saffron amongst foreign merchants (Johnston 1797, p. 287).

141 Turmeric itself may also be adulterated, e.g. by the toxic dye metanil yellow (Dhakal et al., 2016).

142 Including in the Netherlands, Turkey and Morocco (SF, pers. obs.).

143 Rekha et al., 2011; Marieschi et al., 2012; Husaini et al., 2013; Huang et al., 2015; Bosmali et al., 2017; Xiao et al., 2017; Moratalla-López et al., 2019; RASFF, 2019.

144 *Gardenia jasminoides* [Rubiaceae] is grown in China and used as a yellow dye and in traditional medicine (Xiao et al., 2017). It is also an alternative source of crocin (Ibid.).

- Animal-derived products: dried meat fibres and honey.
- Artificial colours: tartrazine (E102), carmoisine (E122), Ponceau 4R/cochineal red A (E124), toluidine red, Sudan 1 and Sudan 4 (in EU specimens).[145] Colourings may be used to dye genuine saffron that has previously been exhausted by infusion, then re-dried.
- Minerals, chemicals and other man-made substances: nylon fibres, glycerine, gypsum, chalk, starch, sugar, glucose, sodium sulphate and even codeine phosphate cough syrup.

Sometimes the fraud may be to deliberately under-dry the saffron, leaving its moisture content too high and so increasing its mass, which in turn adversely affects its keeping qualities and may cause it to go mouldy.[146]

Adulteration and counterfeiting have potential impacts on human health.[147] For example, some artificial dyes are not approved for food use (e.g. Sudan dyes in the EU because of their carcinogenic properties[148]), and undeclared additives in foods present risks to allergy sufferers.[149]

2.8 *Quality Designation and Testing*

No quantitative analytical techniques were available to Hertodt or his contemporaries for discriminating fake and real saffron, or to measure saffron's quality. Various qualitative procedures were detailed by Pereira in the nineteenth century, including saffron's characteristic colour change to blue on the addition of sulphuric acid.[150] Much of today's global saffron trade uses the International Standards Organisation's specification (ISO 3632-1: 2011) and test methods (ISO 3632-2: 2010) for saffron.[151] Tests include moisture and volatile matter content (which would highlight under-dried samples) and UV-visual spectrophotometric measurements of saffron extracts at:

- 440nm, giving the "colouring strength", relating to the amount of crocins present,
- 330nm, relating to safranal content,
- 257nm, relating to the picrocrocin level.[152]

145 In Iran, 52% of 573 samples of saffron-containing dishes from Tehran restaurants tested positive for synthetic food colourings (Moradi-Khatoonabadi et al., 2015).

146 Bradley, 1727, p. 150.

147 Moratalla-López et al., 2019a.

148 "Background about Sudan dyes in food." EU MEMO/05/61 Brussels, 24 February 2005. europa.eu/rapid/press-release_MEMO-05-61_en.pdf.

149 Moratalla-López et al., 2019a.

150 Pereira, 1840, footnotes pp. 675–676.

151 Saffron specification, see https://www.iso.org/obp/ui/#iso:std:iso:3632:-1:ed-2:v1:en Test methods, see https://www.iso.org/obp/ui/#iso:std:iso:3632:-2:ed-2:v1:en.

152 At these wavelengths, the levels of certain other molecules may also inadvertently be

Depending on the results, the saffron is graded in categories I, II or III.[153] Category I has the highest readings and therefore is the most potent saffron gram for gram. A strength of ISO 3632 is that it allows customers to take into account the price per unit of colouring strength rather than simply price per gram when comparing different kinds of saffron and their value-for-money.[154]

However, ISO 3632 is criticised because it cannot fully identify saffron adulteration,[155] as exemplified by a highly sophisticated fraud reported in 2017. Here, a sample of powdered saffron that ISO 3632 had highlighted as impure was demonstrated to be a 100% synthetic mixture of chemicals mimicking both the appearance of powdered saffron and its UV-visible spectrum specific absorbance values.[156]

Alternative test methods are outlined by Moratalla-López et al., who called to amend, or replace, ISO 3632's spectrophotometric analysis with HPLC-DAD (high-performance liquid chromatography with diode-array detection).[157] Additional screening techniques include DNA-based technologies, that can be used to elucidate quickly and cheaply the presence of several common plant-derived adulterants.[158] There are also analytical methods that can discriminate the provenance of saffron samples.[159]

3 Saffron in Orthodox European Medicine after *Crocologia*

3.1 *Changing Perceptions of Saffron's Value*

Saffron was held in high esteem in seventeenth century European medicine and this viewpoint would remain current for almost another 100 years. Not only was saffron praised by physicians and apothecaries, it was also valuable in veterinary medicine; it was "a great Cordial, and of universal Use in Medicine, and is given in many different Forms, but chiefly in Substance or Tincture: and there are few Distempers incident to Man or Beast, wherein it is not of Service,

measured, giving over-estimates of the picrocrocin and safranal levels (Moratalla-López et al., 2019b).

153 A lower grade, IV, is no longer used.

154 However, only the very best artisan producers print their grade and colouring strength on their product packaging, and this important information is missing from "mass-market" brands of saffron (SF, pers. obs.).

155 Bosmali et al., 2017.

156 Ordoudi et al., 2017.

157 Moratalla-López et al., 2019a; 2019b.

158 E.g. Marieschi et al., 2012; Bosmali et al., 2017.

159 Maggi et al., 2011; D'Archivio et al., 2016; Rubert et al., 2016.

both externally and internally."[160] However, less than five decades after this was written, saffron's efficacy was placed under serious doubt by some physicians. Saffron usage entered a long period of decline, leading to its eventual abandonment in conventional medicine.

Attitudes to saffron will be detailed here largely using the UK situation as a typical example of the wider picture in Europe.[161] In Britain, one catalyst for saffron's decline in reputation was the publication in 1768 of Edinburgh physician Dr William Alexander's *Experimental essays*. Alexander self-experimented with increasing doses of various substances, including saffron, "with a view of selecting the valuable from the useless—determining the doses and operation of the former—and throwing the others into that contempt which they justly deserve."[162] After administering himself with 4 scruples of saffron (one sixth of an Apothecaries' ounce), Alexander noted no affect on his temperature, heart rate, nor the colour of his urine or sweat, though the saffron did colour his excrement.[163] By contrast with his saffron experiments, Alexander's near-fatal experience with doses of camphor led him to conclude that saffron was a very weak substance.

William Cullen, professor at the Edinburgh Medical School, publishing some years later, stated that writers on the materia medica had made "manifestly extravagant" reports on saffron's activity, which he felt, based on his own observations with the drug, were unjustifiable.[164] Cullen claimed that he had "hardly" observed any increase in the pulse rate, or an anodyne or antispasmodic effect after administering saffron. Furthermore, Cullen wrote, saffron, when used as an emmenagogue, had "disappointed his expectations." The London College of Physicians had recently removed tincture of saffron from their pharmacopoeia, but did retain saffron in two preparations: the compound tincture of aloes (*tinctura aloes composita*) and pills of aloes with myrrh (*pilulae ex aloes cum myrrha* or Rufus' pills, later called *pilulae aloes et myrrha*). Cullen admitted that he had frequently prepared both of these without saffron, and seen no diminution of the preparations' virtues.

160 Gibson, 1721, p. 17.

161 Saffron was included in catalogues of medicinal plants, and in European pharmacopoeias, from the 16th to 20th centuries, then its medicinal and pharmaceutical use in mainstream medicine disappeared (Bagur et al., 2018). Similarly, in the USA, saffron and tincture of saffron were described in the seventh decennial revision of the *Pharmacopoeia of the United States of America* (National Convention for Revising the Pharmacopoeia, 1893), but saffron is absent from the eighth revision (United States Pharmacopoeial Convention, 1907).

162 Alexander, 1768, Preface p. iii.

163 Alexander, 1768, pp. 88–92.

164 Cullen, 1789, p. 313 et seq.

Alexander's and Cullen's views were not unique: a 1788 commentator on the new edition of the *London Pharmacopoiea* bluntly asked with respect to the modified formula for compound tincture of aloes (or elixir of aloes): "Why is the saffron, a useless and expensive drug, retained?"[165] A similar view was held in farriery: "saffron, though an ingredient in many preparations of the old farriers, is totally undeserving of a place in their materia medica."[166] Although it was formerly considered very active, saffron was steadily becoming "banished from medical practice."[167]

The advent of synthetically-produced chemical drugs meant many traditional medicinal plants were no longer used in mainstream pharmacy by the nineteenth-twentieth century. But with saffron, in the UK an additional factor probably influenced the drug's reputation: domestic saffron production was slowly dying out,[168] and the quality of the saffron in commerce was very questionable. Alexander's experiments were crude and short-term, but were his, Cullen's, and the wider community's observations based on using genuine high quality saffron, or a fake with none of saffron's properties?

The "abuses of the drug-trade and sophistications of indiscriminate compounders of Medicine had become notorious" in the closing years of the eighteenth century, yet surgeons and apothecaries, despite their best efforts, had been unable to rectify the situation.[169] By the early nineteenth century, cake saffron was "a farrago of anything but saffron" disguised with a thin layer of the genuine spice top and bottom. It sold for one third the price to retail chemists and druggists that hay saffron cost the apothecaries.[170] The cake saffron sold in "inferior shops" later became a mere paste of safflower and gum, containing no saffron at all, rolled out thinly and dried in a stove.[171] In Lon-

165 Anon., 1788, p. 196.

166 Boardman, 1805, see under Crocus.

167 Murray, 1824, p. 105. In Germany, similar doubts were expressed about saffron after the publication of Wibmer's observations of the drug's effects (Wibmer, 1831, p. 204).

168 The decline of saffron production in the UK was due to a plethora of factors including high cereal prices during the Napoleonic Wars (1803–1814), rural depopulation because of the Industrial Revolution, the introduction of new flavours into cookery e.g. vanilla and cocoa, and saffron's changing role in medicine. The last saffron grower in England that SF is aware of was in Cambridgeshire, and he stopped production in 1818.

169 Kerrison, 1814, p. 42.

170 Kerrison, 1814, pp. 44 & 45.

171 Pereira, 1840, pp. 674–675. One James Birch was working as a "saffron caker", that is, a manufacturer of cake saffron, in London in 1803 (London Metropolitan Archives CLC/B/192/F/001/MS11936/430/754084), continuing in business until 1818, by which time he was also making "anatta" (annatto) and "sylvester" (an inferior kind of cochineal [OED]) (London

don, this was produced by two makers, and sold for around one-fifth the price of good hay saffron.[172] There must have been a huge temptation, resulting in no little financial gain, to replace real hay saffron with imitation cake saffron.

In North America, "American saffron", namely safflower, was so widely substituted for real saffron that pharmacy students had never seen the genuine drug.[173] A *Canadian pharmaceutical journal* editorial highlighted saffron's and safflower's lack of a close botanical relationship, questioned the inference of similar therapeutic effects between the two substances and denounced the substitution of one by the other. The editorial finished, "though people now-a-days [sic] are sceptical in regard to the medicinal powers of saffron, the remedy has much ancient testimony in its favour, and still retains a place in our pharmacopoeia."[174]

Although saffron still had some devotees, by the mid-Victorian period, the drug was reduced to an antispasmodic "of doubtful value" used in cases of hysteria, and an emmenagogue, again, "of doubtful value," which was not used alone.[175] Tincture of saffron was more commonly just employed as a colorant rather than an active ingredient.[176] It was not only the rich colour of saffron that was valuable in this application, but its special permanence after admixture with other substances.[177] Saffron was also viewed as a "feeble stimulant and diaphoretic" administered to children to encourage the development of certain rashes.[178]

Syrup of saffron, though contained in the earlier London pharmacopoieas, was omitted from the first *British pharmacopoiea* of 1864, and was absent thereafter.[179] Tincture of saffron, by contrast, was retained as late as 1911, though

Metropolitan Archives CLC/B/192/F/001/MS11936/472/940880). That the saffron cakers were based in London, many miles from an ex-farm supply of genuine wet saffron threads, illustrates that it was impossible for them to produce genuine cake saffron.

172 Pereira, 1840, p. 675.

173 Shuttleworth, 1876.

174 Shuttleworth, 1876.

175 Birkbeck Nevins, 1851, pp. 747 & 753.

176 Frazer, 1864, p. 361.

177 Regarding saffron as a pharmaceutical colorant, Birkbeck Nevins (1851) wrote that spirit of nitric ether affected syrup or tincture of saffron less than it did any other colour (p. 121), and saffron (or syrup of roses) was the best colour to combine with alkalis e.g. carbonate of potash (p. 482).

178 Harley, 1876.

179 General Council of Medical Education and Registration of the United Kingdom, 1864. Against this backdrop of diminishing usage, a new basic preparation of saffron was intro-

excluded from the publication by 1917.[180] Two well-known old preparations containing saffron, pills of aloes and myrrh (Rufus' pills) and the compound decoction of aloes (*decoctum aloes compositum*) remained official throughout the nineteenth century. The former preparation, described as a popular and professional purgative in cases of chlorosis and amenorrhea, and the latter, a mild cathartic, tonic, antacid and emmenagogue, needed their saffron "to promote the action of the aloes on the pelvic viscera" according to Birkbeck Nevins.[181] However, by 1899, the formula for pills of aloes and myrrh had been modified, removing the saffron; the compound decoction of aloes formula was similarly altered by 1917.[182] Other official preparations continued in use, but with modified formulae minus saffron e.g. *pulvis cretae aromaticus, tinctura cinchonae composita, tinctura opii ammoniata* and *tinctura rhei*. Saffron was not amongst the materia medica of the 1917 *British pharmacopoeia*.[183] These changes looked like the end for saffron in conventional medicine, but its inclusion in the *British pharmaceutical codex* (BPC), first published in 1907, showed there was still some call for the drug.[184]

The BPC was produced by the Council of the Pharmaceutical Society of Great Britain to supplement the *British pharmacopoeia*, as an authoritative source of information on, and recognised formulae for, medicines that were no longer official, but which were still popular. The 1907 BPC presented 12 formulae containing saffron, including the syrup and tincture of saffron. Demand for saffron still continued, and in 1935, it was one of many drugs for which import duty was suspended.[185] It was still used in some proprietary medicines in the UK in 1943.[186] However, orthodox medicine did eventually abandon saffron.

duced in the 1870s: *glycerinum croci*, or glycerole of saffron (later called glycerin of saffron). Here, saffron was infused in glycerine to make a novel product that did not deteriorate like the syrup or tincture of saffron eventually did. The aroma of the glycerole was said to be stronger than that of the syrup, and the typical copious red-brown precipitate that eventually developed in the syrup of saffron was absent in the glycerole (Anon., 1876).

180 General Medical Council, 1911; 1917.

181 Birkbeck Nevins, 1851, pp. 570 & 221–222.

182 General Medical Council, 1899; 1917.

183 General Medical Council, 1917.

184 Pharmaceutical Society of Great Britain, 1907.

185 Anon., 1935.

186 Council of the Pharmaceutical Society of Great Britain, 1943, see: p. 383, Atkinson & Barker's Infants' Preservative ("For disorders incident to infants and for allaying the pains attendant on the cutting of their teeth, etc."); p. 423, Songo Brand ("Ends all travel sickness. Prevents seasickness, bus, train and motor sickness"); p. 426, Towle's Pink Pills ("For obstinate cases of suppression").

A correspondent to the *British medical journal* in 1950 wrote "*Crocus sativus*, although of great historical interest as the source of saffron, is pharmacologically inert."[187]

3.2 *Current Medical Research on Saffron*

The last two decades have seen a massive resurgence of interest in the medicinal properties of saffron. There is now a significant corpus of peer-reviewed research on saffron, much of it from academic institutions in saffron-producing counties, especially Iran. New research papers, reviews and meta-analyses of previous work are currently being published regularly on the effects of saffron on a range of diseases and conditions, and the preliminary findings are promising.

The usual target for research has been saffron threads, extracts of threads, or purified substances from the threads, e.g. usually crocin or crocetin, but there is some research with safranal.[188] The antioxidant activity of saffron's carotenoids (including crocin an crocetin) is stronger than that of safranal, but the synergistic effects of all saffron's bioactives mean the whole spice has significant antioxidant properties.[189] Studies have been undertaken in animals, cultured cell lines and human subjects, and have been extensively reviewed by different authors.[190] To summarise some of the more recent findings in human trials, using papers featuring meta-analyses where possible:[191]

- **Cardiovascular disease** Saffron or crocin showed a significant effect on diastolic blood pressure, body weight, waist circumference and evidence for a reduction in fasting plasma glucose levels. However clinical trials did not demonstrate as many effects as trials in animals, but this could be because lower doses featured in the clinical trials.[192]

187 Loxton, 1950.

188 Crocin is hydrolysed into crocetin in the intestinal tract, and *trans*-crocetin can cross the blood-brain barrier to reach the central nervous system. It is detectable in human blood serum one hour after administration, reaching a peak four hours after administration. Crocetin forms a weak bond with albumin in the bloodstream, helping its distribution through the body (Bagur et al., 2018).

189 Christodoulou et al., 2015.

190 Lim, 2014, pp. 77–136; Christodoulou et al., 2015; Bagur et al., 2018.

191 Meta-analysis combines the results of a number of (published) clinical trials to increase the statistical power of analysis. Although a powerful approach, one problem is publication bias, since trials giving negative results are often not published and so cannot be included in the analysis (Sun et al., 2018).

192 Pourmasoumi et al., 2019.

- **Depression and anxiety** Saffron or crocin is an efficacious intervention for symptoms of depression.[193] Saffron treatment reduced the symptoms of anxiety compared with placebos.[194]
- **Ocular disease** Saffron or crocin may have positive effects on vision-related parameters in adults with age-related macular degeneration, primary open angle glaucoma and diabetic maculopathy.[195]
- **Fibromyalgia** Saffron had comparable efficacy to the standard treatment, duloxetine.[196]
- **Alzheimer's disease** Saffron has anti-Alzheimer's-disease effects. With human subjects, it improved cognitive function compared with a placebo, and in another trial its effectiveness was comparable to a standard treatment, donepezil.[197]
- **Diabetes** Saffron and crocin have hypoglycaemic effects.[198]
- **ADHD (attention-deficit/hyperactivity disorder)** Saffron had the same efficacy as the drug methylphenidate in children with ADHD.[199]

It is important to stress that the authors of the papers referenced above state that these are preliminary findings from small, essentially pilot, trials. More research is needed, in the shape of large rigorously-designed, long term, clinical trials, to reach firm conclusions.

The possible anti-cancer role of saffron has been investigated in many studies using cell-lines. Saffron has potential anti-tumour effects on leukaemia, lung, breast, skin, prostate, cervical, colorectal, gastric, liver and pancreatic cancers.[200] However, there have been no clinical trials as yet.[201]

Saffron's biochemistry has been the starting point for development of a completely novel class of man-made drug: *trans*-sodium crocetinate (TSC). Originally developed by reacting a saffron aqueous extract with sodium hydroxide (but now synthesised), TSC has better solubility than crocetin.[202] TSC "safely re-oxygenates oxygen-deprived tissue" and "can be used to enhance to cancer-killing power of radiation and chemotherapy".[203] At the time of writing, TSC is in clinical trials for stroke, plus brain and pancreatic cancers.

193 Tóth et al., 2019.

194 Marx et al., 2019.

195 Heitmar et al., 2019.

196 Shakiba et al., 2018.

197 Finley & Gao, 2017; Uddin et al., 2019.

198 Yaribegi et al., 2018.

199 Baziar et al., 2019; Moradzadeh et al., 2019; Naeimi et al., 2019.

200 Patel et al., 2017; Khorasanchi et al., 2018.

201 Leone et al., 2018.

202 US Patent US6060511A; Gainer, 2008.

203 *Diffusion Pharmaceuticals*. https://diffusionpharma.com/product-pipeline/overview-trans-sodium-crocetinate-tsc/ accessed 20/09/2019.

Possibly as a result of published research findings, a plethora of saffron dietary supplements has now been developed, with several trademarked extracts on sale, especially on the internet. Some of these products are targeted to weight loss, some to improving low mood,[204] and others to maintain "normal vision."

4 Conclusion

Crocologia showcases the impressive role of saffron in seventeenth century medicine, and also contains a surprise: the therapeutic uses of other parts of the saffron plant. In the case of saffron tepals and corms, Hertodt lists a intriguing number of applications which are not familiar today. According to him the whole flowers or their preparations have activity against: oedema (it is a specific, he claims), erysipelas, breast inflammation, agrypnia, phrenitis, spleen diseases and insomnia.

In traditional (presumably Iranian, based on the authors' location) medicine, saffron tepals have been used as an antispasmodic, stomachic, anxiolytic, antitumor and antidepressant agent.[205] Ground saffron flowers are recorded to relieve dysentery in Ayurvedi, Unani and Sidha practise.[206] But, current research on tepals has lagged behind that on the threads. The results from recent animal studies suggest saffron tepals have anti-bacterial, antispasmodic, immunomodulatory, anti-tussive, antidepressant, anti-nociceptive, hepatoprotective, renoprotective, anti-hypertensive, anti-diabetic and antioxidant properties.[207]

Hertodt wrote that preparations of the corms can be used to treat oedematous feet, and used against impotence. Corms are not widely employed in traditional systems; a paste of corms is applied to skin affections in Ayurvedic medicine for example.[208] There is a paucity of new research on corms, but a novel glycoconjugate extracted from saffron corms had cytotoxic effects against human cancer cells.[209]

204 One product's marketing information boasts "Move over St John's wort, saffron is poised to revolutionize mood management" (https://www.healthmonthly.co.uk/swanson_superior_herbs_saffron_extract_2_safranal accessed 20/08/19).

205 Hosseini et al., 2018.

206 Quattrochi, 2016, p. 1176.

207 Hosseini et al., 2018; Moratella-López et al., 2019b.

208 Quattrochi, 2016, p. 1176.

209 Escribano et al., 1999.

Several authors e.g. Hosseinzadeah & Nassiri-Asl[210] have analysed historical medical literature and concluded that this is a valuable approach to finding new biological activities (and later, compounds) from various plants. It may be that many other "forgotten" uses for saffron are revealed in *Crocologia*.

210 Hosseinzadeh & Nassiri-Asl (2012) studied Avicenna's (Ibn Sina) the *Canon of Medicine* with respect to saffron. Mollazadeh et al. (2015) looked at the work of Rhazes. Javadi et al. (2013) surveyed a range of other traditional Islamic medicine books.

Notes on Our Translation Methods

For our translation, we initially used a digitised version of *Crocologia*, bearing a Bayerische Staatsbibliothek label, from Google Books in 2012. Later we were able to source a printed-on-demand paper copy, scanned from a different original. Comparisons of both these versions allowed us to decipher letters obscured by ink blots. Translating *Crocologia* presented several challenges: it has highly abbreviated ingredients, authors and book citations; alchemical symbols; and the term *crocus orientalis* when specifying saffron.

Our first task was to search *Crocologia* for fully expanded versions of abbreviations e.g. "*crocus orientalis*" represented by "*croci or.*". Next, officinal names for simples and compound drugs, apparatus, techniques etc. were transcribed into English (where possible) using mainly Mosan (1605), Parkinson (1629, 1640), Brugis (1640), Tomlinson (1657), Culpeper (1683), Salmon (1694), Blancard (1708), Quincy (1719), Philanthropos (1741), Motherby & Wallis (1745) and Barrow (1749). For some highly specific words, just one reference provided a means of translation, and is given under the relevant Glossary entry. When our translation yielded an unlikely-sounding substance e.g. oil of frogs, we checked in contemporary English medical books to verify whether that item was indeed used in pharmacy.

Many of the ingredients in *Crocologia* are unfamiliar nowadays and no longer in commerce, and the source-plants of several were not categorically identified until the nineteenth century. Botanical names for these were from Flückiger & Hanbury (1874), Mabberley (2017) and The Plant Source v. 1.1 website, with the correct authority and the botanical family quoted (so that the relationships between different plants are clearer) in the Glossary entries.[1] We give synonyms in cases where a plant's newer name is not well known.

Alchemical symbols were transcribed as in Salmon (1694) and Blancard (1708) except for the term "*aristol.*⊙.", which is seemingly specific to *Crocologia*. Normally "⊙" was used to signify the sun or gold, but makes no sense in combination with "*aristol.*", which is the plant *aristolochia* or birthwort. We therefore went back to a particular recipe's original source to learn that "*aristol.*⊙." was Hertodt's special abbreviation for *aristolochia rotunda* or round birthwort.

We retained Hertodt's original arrangement of lines in his recipes.

1 In keeping with Mabberley (2017), we have retained the Compositae, Gramineae, Labiateae, Leguminosae and Umbelliferae family names, rather than use the unnecessary (as we see it) modern alternatives: Asteraceae, Poaceae, Lamiaceae, Fabaceae and Apiaceae.

 | DOI:10.1163/9789004435292_003

Crocus orientalis (as used by Hertodt many times) is not a modern or historical *Crocus* species-name. It is usually absent from English-language medical publications of the seventeenth and eighteenth centuries, although Partridge (1682) uses the literal translation "oriental saffron" in his English version of *Thesaurus & armamentarium medico-chymicum* by German author Mynsicht. However, we feel this is misleading to the modern reader because it can wrongly imply the saffron was an import from the Far East, when instead saffron was a well-established European crop when *Crocologia* was written.

We surveyed the literature for *crocus orientalis* and found that its meaning differed over time and between locations. For example, in the USA, *crocus orientalis* was described as a drug from Ceylon (Sri Lanka) that closely resembled saffron.[2] *Crocus orientalis* and *crocus Austriacus* were implied as synonymous in nineteenth century Germany.[3] An earlier German source states *crocus orientalis* is just saffron (versus *crocus occidentalis* or safflower).[4] In the *Institutiones chirurgiae, seu compendiosum Collegium Chirurgicum Universale* of 1737, *crocus orientalis* is Wienerische Saffran ["Viennese saffron"].[5] Meanwhile in France, *crocus orientalis* meant any of a number of autumn or spring flowering *Crocus* species with white, violet or yellow flowers and was not applied to saffron at all.[6] To Spanish readers, *crocus orientalis* was "açafran ordinario" and *crocus hortulanus* was safflower.[7] Contemporary of Hertodt, Johann Schröder, writing soon after *Crocologia*'s publication,[8] defines *crocus orientalis* as "Orientalisch Saffran, safran du Levant, Eastern saffron, Oesterse Saffraen" and *crocus Austriacus* as "Oesterreichische und Ungarische Saffran, safran d'Austriche ou de l'Hongrie, Austrian or Hungarian saffron, Ongerse Saffraen". Hertodt himself states he believes *crocus orientalis* is from Austria (p. 54), and later writes that it is brought from Hungary and Austria (p. 64), yet he still also uses the term *crocus Austriacus* as well as *crocus orientalis*. We have kept both descriptors, and to avoid confusion about, or commitment to, any geographical provenance meant by *crocus orientalis* we render it as "eastern saffron", like Schröder.

There is no reference list or bibliography in *Crocologia*, so we have constructed one. We give the Latinized and original names of authors. We quote the majority of names in their original language in our text except for those that

2 Wood & Bache, 1856, p. 337; Sweringen, 1873, p. 147.
3 Gruner, 1855, p. 98.
4 Zedler, 1793, p. 1680 et seq.
5 1737, p. 184.
6 de Tournefort, 1700, book II, p. 26.
7 Fontecha, 1606, *Diccionario* p. 37.
8 Schröder, 1685, p. 116.

we felt are more well known today in their Latin form e.g. Clusius. To find the full titles for Hertodt's abbreviated references, we searched the WorldCat and CERL databases.

In general we have used a light editorial touch, except in Chapter IV's complex descriptions of mouse traps, and Chapter VII's long instructions for making particular medicines, which were hard to understand in their original form. Our editorial additions elsewhere are in square brackets, including those few incidences where it was difficult to read a quantity for a recipe, or where no quantity was specified.

Regarding the different typefaces in *Crocologia*, Hertodt's original Italic book names have been retained, but we changed his cited author's names into Roman font. All Greek characters have been preserved in our translation, with a transliteration in square brackets, and a translation (where appropriate) in double inverted commas in square brackets. Hertodt's use of Fraktur to quote German terms and for labelling medicines has been retained. However, where he used Fraktur for a book title we have used italics, and for an author's name we used Roman font.

In an attempt to systematise spellings of the many now-obsolete terms throughout our translation, we referred to the Oxford English Dictionary, though several technical words in *Crocologia* are absent from that. Lastly, modern sources (e.g. OED) treat some words in *Crocologia* as synonyms e.g. eclegme and lohoch (see Glossary), but checking historic material (e.g. Tomlinson, 1657), it is clear that there were sometimes subtle differences in meaning which are lost to us today (see Glossary). We have therefore retained (potentially) synonymous words, both to preserve the richness, and specific meaning, of Hertodt's original writing.

Translated Text

FIGURE 1 The Legend of Crocus and Smilax (see p. 46 for an explanation). This image is from a copy of *Crocologia* in the Wellcome Collection London.

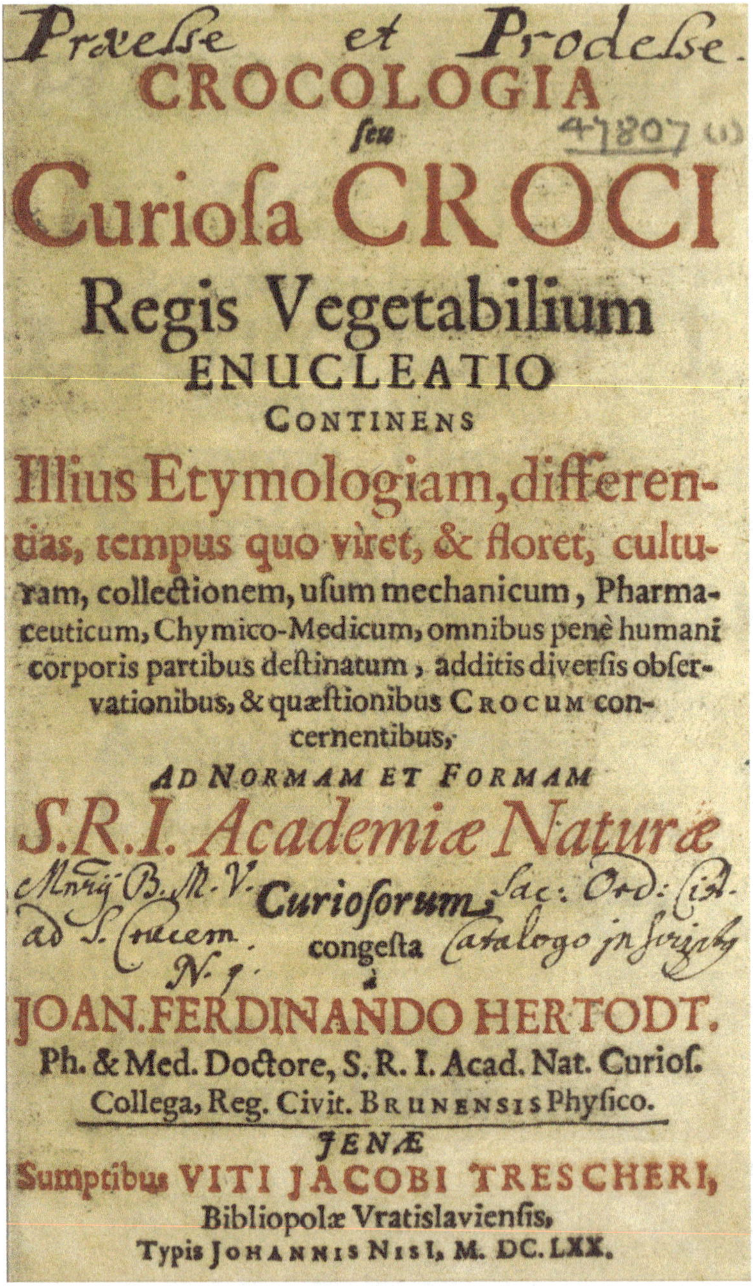

Præesse et Prodesse.

CROCOLOGIA

seu

Curiosa CROCI

Regis Vegetabilium

ENUCLEATIO

CONTINENS

Illius Etymologiam, differentias, tempus quo viret, & floret, culturam, collectionem, usum mechanicum, Pharmaceuticum, Chymico-Medicum, omnibus pené humani corporis partibus destinatum, additis diversis observationibus, & quæstionibus CROCUM concernentibus,

AD NORMAM ET FORMAM

S.R.I. Academiæ Naturæ Curiosorum congesta

à

JOAN. FERDINANDO HERTODT,

Ph. & Med. Doctore, S. R. I. Acad. Nat. Curios. Collega, Reg. Civit. BRUNENSIS Physico.

JENÆ

Sumptibus VITI JACOBI TRESCHERI,

Bibliopolæ Vratislaviensis,

Typis JOHANNIS NISI, M. DC. LXX.

FIGURE 2 The title page of *Crocologia*. This image is from a copy of *Crocologia* in the Wellcome Collection London.

CROCOLOGIA

or

A Detailed STUDY of SAFFRON

the King of Plants

INCLUDING

Its etymology, species,
the time when it emerges & flowers, cultivation
harvesting, mechanical, pharmaceutical and chemical-medicinal uses
Intended for almost every part of the human body,
with the addition of various observations and questions related to
SAFFRON,

ACCORDING TO THE LAWS AND METHOD
of the *Academia Naturae Curiosorum* of the Holy Roman Empire

by

JOHANN FERDINAND HERTODT.
PhD, MD, Fellow of the *Academia Naturae Curiosorum* of the Holy Roman Empire
Physician in the Royal city of Brno.

JENA
Published by VEIT JAKOB TRESCHER,
Breslau Bookseller,
From the Printing House of JOHANN NISE, 1671[1]

1 This is the same publisher and printer as for Hertodt's second major work, the *Opus mirificum sextae diei* (1670).

TO LEOPOLD I,
FOREVER AUGUST,
THE MOST POWERFUL
and
MOST INVINCIBLE
EMPEROR OF THE ROMANS.
KING OF GERMANY, HUNGARY,
BOHEMIA, DALMATIA,
CROATIA, PRUSSIA, etc.,
Archduke of Austria,
Duke of Burgundy, Styria,
Carinthia, Carniola,
&
Württemberg, etc.,
Count of Tyrol, etc.

MOST VENERABLE CAESAR!

Saffron, the king of plants, symbol of noble humility, kneels before the sacred cushions of your feet, with the intention of offering a homage of due reverence and allegiance; and it does not fear rejection. In fact, the flower is solar, and indigenous and born in the Austrian land.[2] This, in order to be able to bloom under the Austrian sun, is certainly in need of your glorious auspices so that it is irradiated more.

O! Austria, your land indeed comes with the blessing of the Lord; not the wet Pleias, not the stormy Orion, not the furious Aeolous will harm it. May the wind of the Austrian favour last long in my garden, and the spices of saffron will grow.

Most great Caesar, receive this humble and little flower in want of a great patron, and protect it in that bosom, where worshipping one's own gods just with milk and wine according to one's own possessions was once considered no offence. The gods themselves (as Pythagoras wrote a long time ago in his *De divinis*) do not look at the abundance of things, but at the intentions of those offering.

When I offer the Austrian flower, undoubtedly I intend one thing figuratively, that, along with saffron, the House of Austria may flourish, bloom and prosper endlessly, and its progeny be master of the gates of their enemies.[3]

This I wish, this I pray

The most humble retainer of Your Holy Imperial Majesty,
Johann Ferdinand Hertodt, PhD & MD

---

2 Saffron has a long history of cultivation in Austria. Abl (1854) writes that saffron was introduced to Austria in 1770, and lists the following places in Lower Austria where is was cultivated: Eggendorf, Kirchberg, Loosdorf, Maissau, Melk, Ravelsbach, Wagram etc. (and in Hungary: Tolna, Heves, Zala, Baranya, and Bihor (now in neighbouring Romania)). However, research by modern-day Austrian grower and botanist, Dr Bernhard Kaar (Director of Wachauer Safran, pers. comm., 14/10/2018) has revealed that saffron production dates from around 1450, and according to legend, from 1197. The last three farmers there stopped growing saffron in 1911, and it was reintroduced by Dr Kaar.

3 House of Austria: the Habsburgs.

To the benevolent and honest reader,

You see this little book under the name of the *Curiosi* come to light, but first, the philosophy of our institution should be made known to you, so that from a slanting supposition you do not believe that it has been filled by the *Curiosi* also of such things which are of no use, such as: who described gigantic statues, or labyrinths, what kind of thread was the famous Ariadne's thread, of what nature was the famous inextricable Gordian knot, what was Midas' wand that transformed everything into gold, in what manner Jupiter did marry Juno, what is the number of the flies covering four horses, what was the famous bronze cauldron from Dodona,[4] what was the weight of the Egyptian Pyramids? These things have nothing but bombastic presumption and are a fatuous and idle display of outlandish grandeur. I present you a flower, which is simple but extraordinary in value: *saffron*. If only you observe it closely, it represents a symbol of your life,[a] where the excellence of the one or the other thing is turned into mockery in the space of one hour. Not unjustly the philosopher Heraclitus of Ephesus, deploring, lamented with regard to the life of all human beings, that there is nothing whole, there is nothing in a human being, which is not transitory, flowing, and deceiving.[b]

However, once you have determined the innermost anatomy of this flower, you will find that it has the greatest medicine against diseases, to preserve life (than which nothing is more precious) and to delay the time of death (if it is permitted to say so).

On this very account, honest reader, collect honey with the bees, and while they settle on a fruitful branch of olive-tree, enjoy this flower along with its effects for a long while, to induce uninterrupted good health.

a Gregorius Pachymerius, *Historiae* Book VIII.
b Laërtius, Book IX. *In vita sua*.

4 Dodona, in NW Greece, was the site of a great oracle to Zeus. There, a sacred oak was surrounded by bronze cauldrons which created a continuous ringing sound. Later these were replaced by a bronze statue holding three chains of knuckle bones, which knocked against a single cauldron in the wind, maintaining a ringing sound.

CABALISTICHON

quo

AVCtorI CroCoLogIae CVrIosVM

Orpheo oCCInIt

Orpheus Evridicen parco revocabat ab orco,

Id seû Alter agis restituendo Crocum.

["CABBALA

for Orpheus,[5] the author of *Crocologia*

Orpheus brought back Eurydice from a short death,

you, the second (Orpheus), do this by returning saffron."]

1	2	3	4	5	6	7	8	9	10	20	30	40	50	60	70
A	B	C	D	E	F	G	H	I	K	L	M	N	O	P	Q

80	90	100	200	300	400	500
R	S	T	V.	X	Y	Z.

O	50	E	5	p	60	c	3	
r	80	u	200	a	1	a	1	
p	60	r	80	r	80	b	2	
h	8	i	9	c	3	a	1	
e	5	d	4	o	50	t	100	
u	200	i	9	r	80	a	1	
s	90	c	3	e	5	b	2	
	493.	e	5	u	200	o	50	
		n	40	o	50	r	80	493
			355.		529.	c	3	355
						o	50	529
							293.	293
								1670.

5 Orpheus is Johann Ferdinand Hertodt. He wrote his contribution to the scientific journal *Miscellanea Medico-Physica-Curiosa* under this pseudonym.

I	9	a	1	a	1	s	90	d	4	
d	4	l	20	g	7	t	100	o	50	
s	90	t	100	i	9	i	9	c	3	
e	5	e	5	s	90	t	100	r	80	
u	200	r	80	r	80	u	200	o	50	308
	308.		206.	e	5	e	5	c	3	206
					192.	n	40	u	200	192
							544.	m	30	544
									420.	420
										1670.

Matthias Franz Hertodt, PhD & MD[6]
renowned for his medical profession in the Margraviate of Moravia

∴

For *Crocologia* by the most esteemed Master HERTODT

Nard does not delight as much with its sweet scent,
And pleasant is the odour which spreads from red roses:
Thus this book offering delightful libations of saffron,
Possesses the pleasant gifts of the salubrious flower.
Egypt brings forth its balsams of the extraordinary plant,
Austria contends (with it) zealously, while producing saffron.
Who could deny that this land, whose fertility nourishes the noble
Seed of saffron, is located under a favourable climate?
Who could deny that Hertodt was born under a lucky star,
Since he is indeed rich in the flowers of intelligence?
Galen gives the skills, public honours are a gift of Baldus,
Whoever has both desires nothing more.
The profession gives Hertodt the skills, now saffron adds the honours,
This is enough, Hertodt; you take two rewards.

With honour by
Jakob Heinrich Kielmann,
PhD & MD, Viennese Society

6 Matthias Franz Hertodt also wrote the preface for his cousin's first book, the *Tartaro-Mastix Moraviae* (1669).

∴

For *Crocologia Curiosa*
by the most excellent, most esteemed and most experienced
Master
JOHANN FERDINAND
HERTODT von Todtenfelt/
ORPHEUS among the *Curiosi*

By no means Aurora spreads on the lands too many *saffron-coloured*
Rays, when she carries the returning days in the chariot:
The fertile Mount Tmolus by no means produces ever so great a number of[7]
Saffron-like spices, and covers the fields with flowers with golden foliage:
Never in Rome more abundant rain wetted the theatres,[a]
When Falernian wines run mixed with *saffron*:
Hertodt's golden offspring pours forth many gifts,
This golden writing is produced with purple *saffron* crocuses.
How many advantages, he says, the threads manifest.
Safflower!
For how many diseases *saffron* could be a remedy!
Who will fear diseases and death in future? It has been discovered,
Thanks to you, who is the powerful Master Lachesis.[b]

Thus,
with the intention of manifesting sympathy to a dearest colleague, with a hastening pen, from the Silesian Breslau applauds

Philip Jakob Sachs von Löwenheim,
PhD & MD, fellow of the *Academia Naturae Curiosorum*.

∴

a In the amphitheatres, the immoderate Romans sprinkled the spectators with wine mixed with saffron, see Martial, *Liber spectaculorum*, epigram 3, book III; epigram 64, book V; epigram 26, book IX; epigram 39; Pliny, *Historia naturalis*, XXI, ch. 6; Seneca, *Moral epistles*, 90; Lipsius, *De amphiteatris*, ch. 16; Brodaeus, *Miscellanea*, book V, ch. 6.

b Hertodt in German Herr Todt, has the same meaning as *Dominus mortis* in Latin (i.e. Master of death).

7 Mount Tmolus, now called Mount Bozdağ, is in İzmir Province in the SW of Turkey.

To
The most excellent, most esteemed and most experienced
MASTER AUTHOR
Who is publishing a *Crocologia* with
Eternal fame of name, a dearest
Friend and colleague.

You investigate thoroughly the nature and the properties of *saffron*, and thence
You show which remedy brings innumerable benefits.
This little flower was born in a field from the blood
Of young Mercury (as an ancient story recounts):
And watered by your sweat, and active toil
It sprouts out in laurel and victory for you.
I wish for Hertodt as many prosperous years,
As the flowers of *saffron* germinating in one Autumn!
And that you may be the Master of Death for a long time, but that Death
May be a master to you later after long years, hail!

In haste,
From the Academy of Vienna
June 20th, 1670

Your Excellency's
most devoted friend

Georg Sebastian Jung,[8]
PhD & MD, Fellow of the
Academia Naturae Curiosorum

8 Georg Sebastian Jung also wrote an approbation for Hertodt's earlier book, *Tartato-Mastix Moraviae* (1669).

CHAPTER I

Etymology and Description of *Saffron*

It is characteristic of ancient names to be consonant with their objects and clearly to signify them. They are like mirrors by which we represent to others what our minds perceive. So the Apostle Paul, in the *Epistle to the Hebrews*, in order to show the excellence of Christ says:

Where do names come from?

> *dedit illi nomen, quod est super omne nomen*
>
> ["He gave Him a name which is above every name"].

Names are derived in various ways from a variety of sources. They are derived either from the great Creator of objects and names Himself at the very moment of creation, when He called the man Adam and gave him the name of the earth, so that his name expressed his nature and substance because he was made of earth. Or they are derived indirectly through the angels as in the name John, which signifies grace. Additionally, some names are created by men: thus, from a fortuitous event, Isaac from a laugh; Jacob from the act of stumbling; or from the prejudgement of a future event, like Cain, Abel; or on account of an omen, like Noah, Benjamin; or from the memory of past events, as Babel, Amon; or from colours, as Rufus from red, Albinus from white [*albus*]; from a condition of the limbs or size of the body, like Simon from flattened nostrils [*simis*], Nausicaä from a hooked nose [*naso*]; or from some singularity of birth, like Posthumous because he was born after the death of his father, or Caesar because he was cut [*caesus*] from the womb of his mother. Similarly names are derived from fortune, like Felix or Bonaventure; from virtues as well as from brotherly love, like Philadelphus; from vices, like Brutus; just as a foolish man was named Nabal which indicates a fool, and because foolishness was in him, he was given the name of a fool. In the same way, Antiochus, because he was liberal in his promises but never delivered, was called δωσιν [dosin] in Greek, *dabo* ["I'll give"] in Latin. Natural philosophers, just as those who carefully examine the rules of nature, give names to their produce according to the nature of each thing: among these examples is *aristolochia* [birthwort] ὅτι ἀρίζῃ ταῖς λόχαῖς [oti arízi taís lóchaís: "because it is best for birth"] i.e. it helps extraordinarily those giving birth; *malva* [mallow], μαλάχη [maláchi], because it soothes the bowels; *raphanus* [radish], ὅτι ῥαδίως φαίε-ται [oti radíos faíetai: "it easily grows"] because it

 | DOI:10.1163/9789004435292_005

appears soon after sowing; *χαμαίδρυς* [chamaídrys: "like a small oak"] which the form of the leaves [of this, the germander] suggests, etc.

Origin of the name Crocus

Thus, the poets in giving expression to the visual loveliness of our crocus flower rightly derived its name from the youth Crocus. As he was pining with tender love for the maiden Smilax, he was turned into a flower with his name because he was unable to bear such love, as Ovid says in *Metamorphoses* 4.283, *Et Crocum in parvos versum cum Smilace flore* ["And Crocus with Smilax were changed into tiny flowers"]. Others say that the young man was killed by Mercury when, during a game of quoits, he was struck on the head by a quoit and a *saffron* flower sprang up from his blood.

Some, like Isidore (book XVII, p. 1254) derive the name crocus [saffron] from the town of Coritium where the flower is abundant, since many things take their name from the places where they are more abundant and thrive best.[1]

This is confirmed by our own plums, commonly called Brunensia, which take their name from the place called Bruna [Brno], a Royal city of Moravia, and are called **die Brünner Pflaumen** ["Brno plums"] in German.[2] These fruits are sometimes suspended from the ceiling so that they are not rotted by the damp atmosphere but are exposed to the air. They are very palatable and an effective laxative especially if taken in the early months, when they are far superior even to manna and cassia.

Others say that the name crocus [saffron] comes from *κρόκη* [króki] which means thread in Greek, since its styles are like threads.

Philologists concluded that it is called crocus [saffron] because it flowers with the frost *Συκρύει* [Sykrýei] in winter.

The name crocus [saffron] may be common with several different things: with the egg yolk or the yellow of the egg; with crocus martis; with the oxide of metals; with the yellow stamens of the white and yellow lily; with safflower; with a certain Indian root about which see Nüremberg, ch. 49, p. 315.[3]

What is *saffron*?

In the present work we mean the little plant, which has five or eight leaves twelve digits or nine inches long, quite narrow, grass-like, marked by a whitish sinew extending longitudinally across their middle, among which a short stalk rises, supporting a single flower, similar to the *Colchicum* flower.[4] In the

1 Coritium: the abandoned ancient Syrian city of Cyrrhus, near the Turkish border Province of Kilis (Graesse et al., 1972).

2 The original spelling erroneously reads: **die brünner Pflaumen**.

3 Turmeric, *Curcuma longa* L. (Zingiberaceae), also called Indian saffron. See Introduction.

4 *Colchicum* (family Colchicaceae: a different family from the saffron crocus) is a genus of around 100 species, the most well known of which is *C. autumnale* L. See Introduction.

middle of the [saffron] flower are three or four threads of a fiery red emitting an intense fragrance, representing the beautiful story of those who love violently.[5] Under them there is a small bulb of variable size, pleasant, round, succulent, vigorous, white on the inside and brown or blackish on the outside, protected by a hairy covering.[6] In Latin this is called *Crocus, Crocum.* The redness of *saffron* threads is greater than that of the clotted blood of Hercules, as in Dioscorides [*De materia medica*], book I, ch. 25; Virgil, *Georgics IV*:

> *Et glaucas salices, casiamque, crocumque rubentem*
>
> ["And grey willows, cassia and red saffron"]

In Greek, [saffron is called] *κρόκος* [krókos], *κρόκων* [krókon]; whence *κρόκωτος* [krókotos] "saffron-coloured", and "saffron-coloured, wheel-shaped, fringed court-dress used by the richest matrons"; *κρόκοτος* [krókotos] "saffron-coloured court-robe"; *κροκωτόφορος* [krókotoforos] "someone who is dressed in saffron yellow". See directly s.v. Johann Altensteig, p. 54.[7]

In others [i.e. other languages], castor, cynomorphus, cresta, crastus, Heraclion, Xenaticus, oricum, jamfaram, etc. In German, **gerechter Saffran** ["just saffron"]/**Orthsaffran/Zammersaffran**. For us it is **Landsaffran**/and especially **Zimmersaffran**. In Dutch, **Safraen** and in Bohemian, **Ssaffran/Jsfran**, in Polish, Szafran. In French, Safran. In Italian, Zaffarano, croco. In Spanish, Azafra, Safra, Azafraon. In Arabic, Sahafaran. In Hebrew, Karkon, confirms a Levite. In Turkish, Zafran.[8]

---

5 Saffron flowers usually have tripartite styles, but flowers with four, five, or even six styles can occur (at decreasing frequency), see Introduction. In Switzerland, the traditional names of Princess, Queen and Empress respectively are given to such flowers.

6 A better term would be "fibrous covering", but Hertodt's original wording is "*capillari involucro*". See also plate of crocus flowers. Hertodt uses the term *bulbus* ["bulb"] but in modern botany that word has a restricted definition that does not apply to saffron. A bulb is a fleshy, highly modified shoot, consisting of thickened leaves or leaf bases, which e.g. in an onion can easily be recognised after cutting in half. A saffron plant, by contrast, grows from a corm, a shortened, thickened underground stem. The word "*cormus*", from which "corm" derives, was first used in the English language in c. 1800 (OED).

7 Altensteig is misspelled as Altenstaig in Hertodt's original text.

8 Erstingium (1770) adds the following names: Aurum vegetabile, Cantacon, Cordiale egregium, Crascum, Crucus, Fedum, Gamphoron, Gracis, Gracum, Haema heraclion, Hortus laetitiae, Medicina tristitiae, Oculus gemmae, Ornamenta terrae, Peritium, Pulmonum amina, Rex vegetabilium, Salesor, Sanguis Herculis, Stomachicum egregium, Sumbi, Uterinum singulare, Xanticum, Xanacticum, Zadrik, Zaffaranum, Zafaran, Zanchiron and Zauda. Lim (2014, pp. 77–78) and Quattrocchi (2016, p. 1176) give exhaustive lists of the names of saffron in different modern languages.

CHAPTER II

Attributes, and Species, of Saffron[1]

After an explanation of the name crocus [saffron] and a description of the plant, we will approach its attributes. An attribute is something which species have in abundance from their stock; thus, Aristotle, *Categories* (*liber predicamentorum*).

Attributes of [different kinds of] *saffron*

An attribute is determined firstly by nature: whence the one is called genuine, the other spurious, or wild.

Secondly by time: for which reason the one is called vernal, the other autumnal.

Thirdly by colour: one is dark-coloured, the other is white, golden yellow, and another is of various colours.

Fourthly by the flower: this kind has many flowers, that kind has two flowers or one single flower, which can be rather small or rather big.

Fifthly by the leaves: the one is broad-leaved, the other narrow-leaved.

Genuine *saffron* includes only two subspecies: one with a purple flower, the other, extremely rare, with a white flower.[2] Wild, or rather spurious, *saffron* instead comprises several subspecies and is distinguished into autumnal and vernal. And the latter is further distinguished into broad-leaved and narrow-leaved.

Autumnal

The first is characterised by a slightly purple flower.

The second by a violet flower.

The third, which is called Constantinopolitanus, is characterised by a white flower, which then becomes slightly purple.[3]

---

1 Historical sources list wild *Crocus* species as kinds of saffron, whereas now we would say the opposite: that saffron is a kind of crocus, and would restrict "saffron" for *C. sativus*.

2 No white-flowered varieties of *Crocus sativus* are known, so presumably the plant being referred to here is a white form of the closely-related species *C. cartwrightianus* Herb. This is sometimes offered for sale as "white saffron" by horticultural companies (SF, pers. obs.).

3 We consulted Tony Goode, UK *Crocus* National Collection holder, to attempt to identify the *Crocus* species listed by Hertodt. Though lack of detail in many of the descriptions

 | DOI:10.1163/9789004435292_006

The fourth, called Pyrenaeus, grows with a dark violet flower.[4]

The fifth, called biflorus, is no different from the common autumnal saffron crocus, except for the fact that it flowers twice; in German this is **Saffran der zwenmahl in Jahr blühet** ["saffron which flowers twice in the year"].[5] *Hortus Eystettensis*, section III, autumnal plants, p. 5.

Broad-Leaved Vernal

It comprises the following species

1. Broad-leaved saffron crocus with completely white flowers. Clusius describes it in this way: the first species of broad-leaved saffron crocus has three, and occasionally four leaves, somewhat wide and short, with two white sinews running underneath across their length, among which the flower grows from a membrane. The flower blooms with a short stalk, has six white petals, which become purple in their lowest part near the stalk, and in the middle contains threads with a three-cleft short style.[6] The root is bulbous, white, covered with a light bark, and provided with whitish filaments at the bottom.

   The root of this species, if smeared on the genitals, elicits frequent and thick urine; it dissipates the swellings in dropsical subjects; drags out

precludes a positive identification in most cases, there were some exceptions. The name *Constantinopolitanus* in Hertodt's list is seen as a synonym of *Crocus pulchellus* Herb. (Goode, pers. comm., 18/5/2017; Maw, 1886, p. 38 & 316).

4 *Crocus nudiflorus* Sm., widely distributed in the Pyrenees and long naturalised in UK, being brought here by Knights of St John and others (Goode, pers. comm.; Maw, 1886, p. 38; Blacksmith, 2011). It was first hypothesised that *C. nudiflorus* was introduced to the UK as an alternative source of saffron by Crump (1899), though its stigma is not tripartite like saffron's, and its fresh flowers and experimentally-dried stigmata are low in aroma compared with those of true saffron (SF, pers. obs).

5 No *Crocus* species flowers twice in the year, so possibly this observation was of a mixed population of different species. Brian Mathew states in his seminal work on the genus that the name *Crocus biflorus* Mill. was first used in 1768 in Phillip Miller's *Gardeners Dictionary*. It covers a vast array of species in the current understanding of the genus. There is one in this group which flowers in autumn, the others are spring flowering. Autumn flowering "*biflorus*" is *Crocus melantherus* Bioss. & Orph. ex Maw, a localised species with black anthers. It is not known whether this plant was recognised at the time *Crocologia* was written [*C. melantherus* was used for the first time as a scientific name in 1881 in the *Gardeners' Chronicle* ii, p. 749, though this species could have been recognised before that] (Goode, pers. comm.).

6 Hertodt uses the word "petals", as do many modern writers. The correct modern botanical term for these floral parts in *Crocus* flowers is "tepals" (a word first used in English in 1840 (OED)). See Glossary.

thorns; drives away lice; makes the complexion beautiful; increases and strengthens the fierceness of power in the limbs; restores the colour of dirty clothes.[7] Johann Bauhin, *Historia plantarum universalis*, p. 642.

2. Broad-leaved saffron crocus with striped white flowers.
3. Broad-leaved saffron crocus with an extremely large purple flower.
4. Broad-leaved saffron crocus with a big flower with purple stripes. It regularly grows several leaves from the bulbous root, which are somewhat bent backwards, nine-inches long or longer, narrower at the stalk, slightly stiff, having shoots at the extreme end, and crossed by a whitish furrow on the opposite side, with green edges. A bright-green crest corresponds to the furrow on the opposite side, and the channelled flanks descending from this crest are almost white. Among the leaves, rather large stalks the length of a hand, delicate, feeble, mostly white, wrapped in a white membranaceous sheath, and distinguished by purple and white stripes in the top part, emerge directly from the root. Each stalk carries a sweet-scented flower, which, before unfurling, resembles a simple spike, with six one-inch-long pointed petals. Three of the outermost petals hide in their embrace those staying inside the closed flower and are wonderfully variegated in colour (they are white on the inside and present purple stripes turning into a light yellow on the outside); the three remaining petals, instead, are of a softer violet-colour both on the inside and on the outside, except for a spot of deep violet visible on the external part. In the middle there are three tongue-shaped threads sprinkled with reddish yellow powder, which, full of saffron pigment, surround just as many stamens. Johann Bauhin, *Historia plantarum universalis*, p. 643.[8]
5. Broad-leaved vernal saffron crocus with multicoloured white flower in Clusius; with a white flower with dark blue edges on the side of the petal in Lobel, which is observed sometimes single, sometimes double.
6. Broad-leaved vernal saffron crocus with multicoloured purple flower, described by Clusius, and by Johann Bauhin in the *Historia plantarum universalis*, p. 645.

7 Harrison, 1587, p. 393, wrote that an infusion of saffron "root" in wine has some similar properties and would provoke urination and expel kidney/bladder stones.

8 Much of this description suggests *Crocus imperati* Ten. (from Italy) which has biscuit-coloured yellow outer petals with purple stripes, but lilac blue inside (not white as described here), the inner whorl of petals being the same lilac throughout. The reference to bent back leaves is typical of *C. imperati* (Goode, pers. comm.).

7. Broad-leaved vernal saffron crocus with golden flower, or *calendulatus*, or as described in Johann Bauhin, *Historia plantarum universalis*, p. 644.[9]
8. Broad-leaved vernal saffron crocus with pale flower, ibid. p. 644.
9. Multi-flowered vernal saffron crocus with several white flowers arranged in groups, from the outer top of the outermost whorl of which purple-blue lines are drawn upwards; in German **weißer mit vielen Blumen tragender gestreimbter frühlings Saffran** ["white striped spring saffron with many flowers"]. *Hortus Eystettensis*, images of winter plants, section I, p. 3.
10. Broad-leaved multi-flowered vernal saffron crocus blooms with purple-blue flowers, on whose lowest part white lines stretching out to the edges spring up like rays. It grows with black to ash coloured bulbous roots; in German **Veilbraun gestreimbter frühlings Saffran** ["violet striped spring saffron"]. *Hortus Eystettensis*, winter plants, section I, p. 3. Clusius, book II, p. 203. Lobel, *Observationes*, p. 65, *Adversaria*, 53; *Lugdunum*, p. 1534.
11. Precocious vernal saffron crocus. Its flower is golden on the inside, and presents dark-purple stripes on the outside, its root is oblong and of the bulbous kind, like the other saffron plants; in German **goldgelber gestrimbter frühlings Saffran** ["gold-yellow striped spring saffron"]. *Hortus Eystettensis*, images of winter plants, section I, p. 2; Tabernaemontanus, book II, p. 334; Dodoens, p. 214; Clusius, book II, p. 206; Lobel, *Observationes*, p. 46, *Adversaria*, part a, p. 497.[10]
12. Multi-flowered vernal saffron crocus with very large leaves lying low from the saffron-yellow, ash-coloured bulb, like a sheath on the waxy tawny colour of the flowers, rises up in groups with the corolla totally upright; in German **gelber frühlings Saffran mit viele Blumen** ["yellow spring saffron with many flowers"], from the often quoted *Hortus Eystettensis*, images of winter plants, section I, p. 2; Lobel, *Adversaria*, p. 497.[11]

---

9 Probably *Crocus flavus* Weston [syn. *C. aureus* Sm.] (Goode, pers. comm.).

10 *Crocus angustifolius* Weston (Cloth of Gold crocus) fits this description well (Goode, pers. comm.).

11 This description might also refer to *Crocus flavus* if "tawny" can be taken for a yellow shade (Goode, pers. comm.).

Narrow-Leaved Vernal Saffron Crocus

Subclassified into:

1. Narrow-leaved vernal saffron crocus with a big white flower. In German **frühlings Saffran mit großer weisser Blum** ["spring saffron with large white flower"].
2. Narrow-leaved vernal saffron crocus with a small white flower. In German **frühlings Saffran mit kleiner Blum** ["spring saffron with small flower"]. *Hortus Eystettensis*, section II, p. 3; Dodoens, p. 215.
3. Wild narrow-leaved Alpine vernal saffron crocus with seven and eight leaves, and with bright-violet flowers, in German **Purpurbraun frühlings Saffran** ["purple-brown spring saffron"]. *Hortus Eystettensis*, section II, p. 3; Tabernaemontanus, book IV, p. 334, Lobel, *Observationes* 68; *Lugdunum*, p. 1535.[12]
4. Narrow-leaved vernal saffron crocus with variegated white flower.
5. Narrow-leaved vernal saffron crocus with variegated purple flower.
6. Narrow-leaved vernal saffron crocus with reddish-yellow [orange] flower.
7. Narrow-leaved vernal saffron crocus with variegated flower.
8. Spanish narrow-leaved vernal saffron crocus with dark purple flower, both in Clusius and Johann Bauhin, *Historia plantarum universalis*, p. 644.[13]
9. Smaller Spanish narrow-leaved vernal saffron crocus with purple flower, also in Clusius, and Johann Bauhin, *Historia plantarum universalis*, p. 644.

Finally, although the most esteemed Dr Georg Christoph Petri, our Achilles, in his *Asylum languentium*, ch. 3, p. 45, classifies *Crocus Saracenicus* among thistles because of its prickles, we, however, rightly classify it among the saffron crocuses because of the saffron tint of its flower: whence in Greek it is called *κρόκος* [krókos] and in other languages, wild saffron.[14]

---

12 This is *Crocus vernus albiflorus* [this is a syn. of *C. caeruleus* Weston], which occurs in white and blue forms. This reference to "alpine" suggests quite strongly that this is the species referred to here. Numbers 4–7 below might also be *C. vernus albiflorus* but the descriptions lack useful detail (Goode, pers. comm.).

13 The Spanish crocuses, numbers 8 & 9 in Hertodt's list are highly likely to be *Crocus vernus* (L.) Hill in one of its many guises. The much rarer and distinctly different *Crocus carpetanus* Boiss. & Reut. and *Crocus nevadensis* Amo & Campo occur in the south too, but have distinctive features that should have drawn comment. They have much narrower distributions than the Pyrenean *Crocus vernus* forms (which include white flowered plants) (Goode, pers. comm.).

14 This plant is safflower, *Carthamus tinctorius* L. Petri was quite correct: safflower is not botanically related to saffron. See Introduction.

Its stalk is pale green, round, retuse, sometimes longer than one cubit, divided from the middle in branches. These together with the stalk are in turns closely surrounded by leaves, which are green, hard, ending in a sharp point, and fortified with harmless prickles on the edge; the extremity of the branches is made of scales placed one against the other in the manner of a hollow-tile, and have conical ends, which spread out above the saffron-coloured stamened flowers inclining to red; the seed (because it is sometimes used to feed parrots, is called parrot's seed) is white, strongly angled, twice smaller than the seed of the Peruvian chrysanthemums, with a rough shell on the outside and filled with a cleansing white pulp inside.[15] The root is scattered with filaments, annual and empty. In German this is called **wilter Saffran** ["wild saffron"]. *Hortus Eystettensis*, summer plants, section II, p. 3; Camerarius, *Epitome*, p. 99; Dodoens, p. 358; Lobel, *Observationes*, p. 488, *Adversaria*, p. 374; Clusius, book V, p. 153; *Lugdunum*, p. 1449; Cesalpino, book I, ch. 46, p. 532; Bock, ch. 103, book II; Fuchs, ch. 156; Caspar Bauhin, p. 888; Durante, p. 191.

Safflower growing in mountainous regions which are rich in gold mines is said to attract a solar spirit of such a kind that dyes tin and silver, as if they were gilded. Hiebner, *Mysterium sigillorum, herbarum et lapidum*. To this we can add the wild crocus with dark-blue flower, in German **blau wilter Saffran** ["blue wild saffron"]. See *Hortus Eystettensis*, images of summer plants, section II, p. 5; Tabernaemontanus, p. 372, book II; Lobel, *Observationes*, p. 488; Clusius, p. 152; *Lugdunum*, p. 1449; Camerarius, *Hortus medicus*, p. 35.

15 Peruvian chrysanthemum: an old name for the sunflower, *Helianthus annuus* L. (Compositae).

CHAPTER III

Native Place, and Time When [Saffron] Emerges and Flowers

Native place of *saffron*

Aristotle (*De Adm. Aud.* no. 32) recounts that in Lipara there is a place in which, if someone hides a pot, anything put into it boils.[1] Now, I may say the same about our *saffron*: indeed, in whatever place you plant it, you will see it emerge, as it does not fear Sirius, nor is it terrified at the north wind, when not only the southern countries, Africa, Syrena, Sicily, and France, but also the northern regions, England and Ireland, cultivate it.[2]

Mattioli writes that originally it claimed for itself a place in the city of L'Aquila in Italy. However, our *eastern saffron* (Austrian, I believe) by far precedes that: nowhere is it more vigorous than on Viennese soil, see Dioscorides, book I, *De materia medica*, ch. 25, p. 20. Mattioli, p. 17, and other botanists agree with him.[3]

The leaves emerge in winter, in January, and in spring.[4] Sometimes they last the year through, but they do not wait for the summer.[5] Lobel, *De stirpium adversaria*, p. 53; Tabernaemontanus, book II, p. 334; Mattioli, book I, p. 18.

Saffron blooms in the autumn before the leaves [emerge]; its flowers last for one month, sometimes longer. Erasmus Finx, *Ost- und West-Indischen lustgarten*, p. 482; Mattioli, p. 18.

---

1 Presumably, the abbreviation *De Adm. Aud.* stands for *De mirabilibus auscultationibus*, although it is entry no. 34 and not 32 which recounts about Lipara. Lipara, now called Lipari, is the largest of the Aeolian Islands, off the north coast of Sicily. It is a volcanic area.

2 Sirius was considered the precursor of unpleasantly hot and sultry summer weather (originally named in Greece): the dog days. "Syrena" used by Hertodt could mean Cyrene, an ancient city on the north coast of Libya. Ireland's saffron-growing history is now largely forgotten but Irish saffron was still being imported into England in the early nineteenth century (Mortimer, 1810, entry under Europe: England: Saffron).

3 For the reasoning behind our translation of eastern saffron from *crocus orientalis* see p. 30.

4 This is incorrect: the leaves emerge during, or shortly after, flowering in autumn.

5 That the saffron leaves "do not wait for summer" is a reference to saffron's summer dormancy. Under UK growing conditions, saffron leaves typically die back in April or May, and the saffron field is bare until late September when the plants re-emerge (SF pers. obs.). Saffron's leaves never "last the year through".

 | DOI:10.1163/9789004435292_007

Theophrastus states: *saffron*, both montane and cultivated, begins to bloom in the autumn. And again: it flowers late, and it emerges late, or too soon; either way it has observed the season.

The *saffron* of my cultivation of the past year 1669 started to flower on September 20th, and it lasted for the whole September and October; its 2037 threads, already dried, weighed one avoirdupois unit.

As soon as the *saffron* [crocus] grows flowers, the threads in the middle of the flowers must be removed, at dawn or at dusk.[6] While being collected, the threads must be placed either on papyrus or on a smooth linen cloth; then, they must be cleaned from any dirt. They must be stored in a salubrious place, free from wind, dust, and sunlight. Lauremberg, *Apparatus plantarius*, book I, ch. 6, p. 46.[7]

---

6 Usually growers are keen to collect the flowers as early as possible each day, to (a) preserve the saffron's quality, (b) lessen the chances of the flowers being damaged by birds or bad weather, and (c) to allow time for removing the threads and drying them. SF is not aware of harvesting at dusk being carried out anywhere today.

7 Hertodt gives no description of drying methods (see Introduction), but does mention saffron "dried in the shade" in his recipe for magistery of saffron, p. 78. The traditional drying method used in Austria was to suspend the saffron on sieves over glowing embers (Dr Bernard Kaar, Wachauer Safran, pers. comm., 14/10/2018).

CHAPTER IV

Cultivation of *Saffron*

Werner Rolfinck, *De vegetabilibus in genere*, p. 49.

Saffron is a symbol of perseverance and patience; this little plant, although extremely delicate, is however fertile, and extremely steady. The Ancients said that, before sprouting out in flower, if in some measure trodden and trampled upon with feet, not only does *saffron* not wither, but afterwards it grows more beautifully. And they choose places trodden by the footsteps of men, next to springs [for growing saffron]; from this custom they also call the places which were beaten by the feet of dancers, *κρότηζα* [krótiza]. But our cultivation is carried out in the opposite manner to the cultivation of the Ancients: they [nowadays] believe that nothing is so harmful in domestic economy as extensively treading under foot soil planted with *saffron*, and for this reason they protect the soil as much as possible from the footsteps of men and animals.

They choose a soil which is neither damp nor compacted, and not too manured, but friable, ploughed, exposed to the sun, free from the flow of waters, free from weeds and pebbles; and they consider the best soil to be that where previously planted onions have been already harvested. Pietro de' Crescenzi, *De agricultura*; Coler; Peter Lauremberg, *Apparatus plantarius*, book I, ch. 6, p. 46; Erasmus Finx, *Ost- und West-Indischen lustgarten*, p. 482.

Propagation of *saffron*

The *saffron* [crocus] is propagated very conveniently through bulbs, thanks to which every year it abounds in new offspring: in fact, the growth from its seed is slower.[1] These bulbs must be removed from the soil once they have shed their blossoms; this happens around autumn time.[2] They must be left for half a month, or slightly longer, in a room, not covered with soil and away from the sun, so that they are not heated, but ripen fully. This can be understood from the leaves, if they turn out dry. Afterwards, they must be planted again in furrows equal to an interval of one palm, with the larger

1 True saffron is a sterile hybrid. The saffron lookalike, and also its ancestor species, *Crocus cartwrightianus* Herb. produces viable seed, and could have been cultivated sometimes instead of *C. sativus*, or in mixed populations with it. Douglas (1722) gives a description of the "seed vessel" in saffron plants in England, but acknowledges that the seeds "seldom come to perfection".

2 Saffron corms are lifted at varying times around the world due to climatic conditions and local tradition, but to lift them immediately after flowering would be poor practise. It would deny the plants several months' worth of growth and photosynthesis that would have built reserves for the following year's flowers.

 | DOI:10.1163/9789004435292_008

bulbs placed asunder from the smaller ones. Peter Lauremberg, *Apparatus plantarius*, ch. 6, p. 46.

Some collect the little onions [*caepulas*] of *saffron* in March, and they leave them in the air until the autumn; then they replant them, and they expect flowers the following year.

Others, instead, collect the bulbs in April and May (as we read in De Crescenzi) and they put them in a heap, to clean them and to make them ripen until the autumn; and then they plant them one span from one another in prepared beds, sinking them four digits deep. Conrad von Heresbach, *De re rustica*, book II, p. 209, maintains that the saffron bulb is planted in March.

In certain places, in the seventh or eighth year it [the saffron] is moved to a well ploughed place and thus it is called degenerating.[3]

In the Rhine region it is uprooted in the third year, and dried up until the autumn; then, after removing the outer skin, it is put back with the individual heads distributed half a foot into the ground.

Moles and mice, the enemies of *saffron*

Even in a place of enjoyment, the most perfect peace is a stranger. When mice and moles attack the *saffron* bulbs because of their sweetness, causing extreme damage, in order to extirpate and destroy hidden enemies of such kind, the ancient writers left us a remedy which is extremely effective, especially against moles.[4]

Remedy against moles

Pound white hellebore, and rind of dog's mercury, sift both together with pearl-barley, add eggs with wine and milk, make several small pieces and put them in the holes of the moles.

Giovanni Battista Ferrari of Siena, *De florum cultura*, book III, ch. 3.

The Greeks fought out these enemies living in holes in a simpler way, with fumigation, and were successful with the smoke. They took out the kernel from a nut, or from any other fruit of similar hardness, perforated both sides, and replaced the kernel with chaff, cedar resin, or wax, and sulphur. In the same practice they employed a small vessel to obstruct the scanty air holes, so that they could not admit air. They placed the nut, or any other device for fumigation, kindled, in the widest access to the hole so that the air current received from the outside spreads far within the innermost holes, and,

3 Hertodt uses the word *degenerans* ["degenerating"], though regenerating would be a much better term.

4 Moles and mice are still serious pests of saffron fields. Moles typically make tunnels under the rows of plants, not damaging them *per se*, but making an air gap through which the plants cannot root. Mice then move into the moles' tunnels and eat the corms from below. If the corms do not die from diseases that enter the wounds caused by the mice, they often have insufficient resources to produce new shoots the following year (SF, pers. obs.).

with the underground passages completely filled with dark smoke, the moles escaped immediately or were suffocated.

Others send away these underground enemies with weapons in the following way.

When they realize that the earth is raised and arched over, they immediately go there with stealthy steps, so that the enemy does not run away because of the noise, and they dig out the underground devastator of the earth with a double mattock, or with the pointed part of a hoe or a rake, destroy the underground passages, and paralyse the little beast.

Others bury a clay pot, two or one and a half palms high, in a place exposed to moles, near the surface of the ground or slightly below, and they put a living mole in this very pot. The mole, when, unable to endure the daylight, cannot dig out a refuge because of the hardness of the vessel, nor escape anywhere because of the pot's height and smoothness, complains of its misfortune with a subtle grunting and attracts the neighbouring moles from the inaccessible holes for pity and help. These, blind, crawling into an amusing trap, fall one over the other, and, in the cavity of the pot, they do not stop singing a tragic symphony before they close their eyes in the eternal night of a liberating death. They get caught mostly when it is windy, because they create passages more often in a similar weather.

Remedies against mice

Against mice (which peasants distinguish as country mice and house mice): Constantine Caesar killed house mice with pearl-barley, seed of wild cucumber, black hellebore, colocynth, or even just with hellebore. He filled their burrows with oak-cinders, in order to cause hurtful itching to those strewn by it, and destroy them.

Others mix iron dust with yeast, and when the mice eat it, they die.

Others kill them with blackberry root mixed with bread, butter and cheese.

Some fumigate their burrows with vitriol, wild marjoram, parsley seed, and Roman coriander, or otherwise with bloodstone, or green tamarisk, and thus the mice run away.

Others leave a skinned mouse in the burrows, and thus the other mice, fearing that this might happen to them too, bid farewell to their little houses.

Others catch house mice with thick oil dregs, which they pour in a copper basin and place in the middle of the house. In this way all mice, both those living there and those from nearby places, converge there, and stick to it.

Pliny exterminates country mice with ground elecampane; with carline thistle, with the addition of water and oil, he draws them out and kills them at once, if they do not assume water immediately.

Others kill mice with orpiment, or arsenic. But since these are poisons, the mice tasting them can easily poison other foods with their contagious touch and breath.

A safer remedy is made out of fragmented glass, or fresh white lime plaster scattered in flour, and kneaded with cheese.

But in addition to these, cats are good too, as well as mouse-traps, which can be built in this way.[5]

The majority of mouse-traps are put together in a quadrangular shape with beech boards carefully worked in the way of a small box or a longish little house. One of the little boards, put in the place of the floor, supports the other three standing in the way of walls, and is in some measure prominent in position. The other board, laid on top in the likeness of a ceiling, extends moderately to make eaves. The little entrance is closed by a portcullis, or a swinging door, which inserts into twin iron rings on both sides onto very thin iron pins and is raised by a rather long bar, like a swing-beam. This bar, thicker in the part where it is connected to the portcullis, gradually tapers to a point. In the middle of the top board there is a two-pronged wedge, which intercepts the bar crossing over with an iron wire so that the moving bar pushes the portcullis up and down. On the same top board under the extreme point of the bar, an opening allows an iron wire to pass through, which is wound around the chain of the little iron wire lying across the opening, and bent backwards in the top part, shaped like a hook in the bottom part. As the top hook is fixed to the top bar which easily lifts the portcullis, the bottom hook holds fast the bait. When a little mouse, creeping along, greedily bites the bait, the top chain being withdrawn, it causes the portcullis to be suddenly cast down. Guilty of heedless gluttony, the little mouse turns the eating place into its own prison, and is soon going to suffer a droll death by a jovial and playful executioner, the cat.

The following trap is simpler but deadlier. A single rather wide board, fastened from the back, is furnished with a small opening large enough to fit a mouse-head. An insidious square space, not very big, opens rather inwards, moving away from the front board of the mouse-trap, which leans askance. The bottom edge of the opening is lined with three slightly protruding iron spikes. Likewise, a further iron rod, bent backwards on one side, is attached

5 Hertodt's fierce contemporary critic, Ardensbach, thought it "not seemly" for a physician like Hertodt to instruct on methods of controlling mice and making mouse-traps, see Appendix 2.

to the top board pointing towards the bottom board. The other side of this rod lies across the abovementioned spikes, stretching out in an independent line, whose extreme end finishes slightly beyond the opening, and is fastened to the board with bent rods like a lattice. A slender rope connects a small wooden wedge to the rod, which is violently pulled to the top of the iron line. One end of the wedge, bound with a thin string, is placed on the top board, the other placed under with a hook very similar to the one that we have described in the previous mouse-trap. This too, hooked at the bottom, presents a deceptive bait, and as soon as the bait is even slightly touched with greedy tooth or moved, the wedge suddenly falling down unlooses the rod from the hook. The rod, falling down with great violence, cuts off the head of the glutton, and pierces it with the underlying spikes; and, as above, greediness ends in a sad feast. Giovanni Battista Ferrari of Siena, *De florum cultura*, book 1, ch. 3, p. 267 ff. where he also adds images of instruments for mice.

Secret solution for the multiplication of produce

The clear desire to be of service to society urges me to add, at the end of this chapter on the cultivation of *saffron*, a singular multiplicative solution known to few. If seeds of any kind are soaked into it in the way explained below, and then dried in the shade, and then sown when the moon is waning, they show extraordinary effects indeed.

Firstly, in this manner the seeds are free from worms.

Secondly, the field, except for the usual ploughing, does not need manuring.

Thirdly, if sown with such seed, a field is ready every year for propagation.

Fourthly, half of the seeds prepared in this manner are enough for propagation.

Fifthly, the field produces double the produce.

Sixthly, if a small piece of burnt wood, of the size of a very small finger, is immersed in this solution, and if this is applied to the root of any plant, it wonderfully protects the root from all insects.

This multiplicative solution is thus prepared:

Take of: Spring water, 13 Moravian measures;[6] add salt of nitre not yet completely purified, 3 ounces; shavings of ox or goat horn, 3 ounces; chimney-soot, 6 ounces; ox dung, quantity excreted at one time; hens', or doves' dung, 1 handful; horse dung, 1 handful; and one head of garlic. Mix all these things and put them in a bronze vessel containing the abovementioned

6 Moravian measure: see Glossary for suggestions as to what this is.

water, let them boil up for one hour, and once the solution has cooled down, filter it through a loosely textured cloth, and store it for use.

Then take a large vessel, afterwards evenly place there whatever seeds you intend to sow, so that the one does not stand out over the other. Straightaway mark the vessel with charcoal one palm [width] above the seeds thus distributed, pouring over them the amount of multiplicative solution which reaches the marked spot. Let the seeds soak in the vessel covered with cloths for 24 hours, or for 36 hours if they are seeds of pulses. Afterwards, take out the seeds thus prepared, and diligently let them dry in the shade, to preserve them for further use.

CHAPTER V

Selection of *Saffron*

He who takes heed that he may not be deceived, is hardly on his guard even when he is on his guard. Even when he has thought that he was on guard, often this wary man is caught. Plautus, *Captivi*. If this was ever true, it is especially true nowadays, when trust is absent and deceit rules. I am silent about many things, but I [do] speak out about *saffron*, that regal flower. In what manner is it usually adulterated with various dyes? When it is mixed with litharge. One tricks the buyer simply with ground residue of *saffron*; [another] one cheats by increasing the number of threads with the addition of fibres of salted meat. In order to make it weightier, some people hide *saffron* in damp underground places;[1] others contaminate it with the flowers of safflower (which is extremely common).[2] You will understand that the *saffron* is genuine and free from any dye, if, when it is kept in a suitable location of the house, it continues to be extremely fragrant in its vigour, fresh, and of a beautiful colour, with a little bit of white in the tendril, oblong, uniform in all its parts, not perishable. It is full, nothing has been removed from it, it tinges the hand when wetted, it is somewhat sharp, it does not smell of mould nor rots. Dioscorides, book I, ch. 25, p. 1909. Conrad von Heresbach, *De re rustica*, book II, p. 209, agrees with the latter, stating: the best *saffron* must be fresh, fragrant in its scent, it must stain the hand a golden colour on rubbing it, it must be oblong, slow in the breaking.[3] Ruel, *De natura stirpium*, book II, ch. 11, p. 294, adds that the best *saffron* is very thick, with short threads, and, when the handful is brought to the mouth, it stings the face and eyes a little.

If it lacks these qualities, then it is adulterated.

In fact, when it looks dusty, then suspect that it has been mixed with ground litharge or residue of *saffron*. When must has been added, its flavour is sweet. Thus, if it has been left in storage in damp places, its colour appears dark, and it smells of mould. However, only the odour will determine if it is safflower, or if the numbers of threads has been increased by adding salted fibres of meat.[4]

1 Saffron is hygroscopic (SF, pers. ob.), thus increasing its weight if stored under damp conditions, but at the same time making it liable to go mouldy.

2 See Introduction for more detail.

3 In other words, not brittle.

4 A nineteenth century test was to rub a sample on a piece of white paper with a moistened

 | DOI:10.1163/9789004435292_009

The native place of the *saffron* should also be taken into account in its selection.

Thus, the Ancients assigned the first place to the Corycian *saffron* of Cilicia:[5] there is a Corycian cave, once inhabited by nymphs, who are called *nimphae Corycides* after it. See Ovid, *Metamorphoses*.

The Corycian *saffron* crocus is remembered by Martial as well:[6]

> *Ultima Corycio, quae cadit aura Croco.*
>
> ["Last effluence that falls from the Corycian saffron"]

They allotted the second place to the Lycian [saffron], plucked from Mount Olympus.[7] Then, the one brought by the Aegeans from the city of Actolia.[8] They judged the Cyrenaic [saffron] and all [the saffron] from the town of Centorbi in Sicily very weak, although they considered it extremely suitable for dyeing because of the abundance of sap and the superior quality of the colour.[9]

---

finger: saffron would give an "intense orange-yellow stain" whereas safflower only made a "slightly yellow mark" (Pereira, 1840, p. 675). Additionally, one could rehydrate the sample in hot water and examine it by eye or microscope to see the original morphology of the constituent different elements (Ibid.). The presence of meat fibres was shown by a characteristic smell after throwing a sample onto hot embers (Duncan, 1808, p. 246).

5 Cilicia: A Roman province the south-eastern part of Asia Minor, roughly equivalent to the modern Turkish region of Çukurova (comprising the four modern Turkish provinces of Mersin, Adana, Osmaniye and Hatay). Corycus (Corycos, Korykos), later called Korghoz (Maw, 1886, p. 60), and now Kızkalesi, is a coastal town in Mersin Province. The Corycian cave inland of Corycus, named by Strabo (*Geography* book XIV, ch. 5, 5) as the source of the best saffron, is a large sink-hole in the Taurus Mountains. The Corycian cave housing the mythological nymphs is elsewhere, in Greece, on the slopes of Mount Parnassus (Strabo, *Geography* book IX, ch. 3, 1).

6 Martial, book XI, ep. VIII.

7 Not Mount Olympus in Greece, but from around the ruined ancient city of Olympos in Lycia, near the southern coast of Turkey. It may or may not be significant that a rare indigenous wild crocus grows in this region: *Crocus wattiorum* (B. Mathew) B. Mathew (Aykurt et al., 2015). This autumn-flowering plant has tripartite red style branches, though smaller than those in *C. sativus* (Rukšāns, 2017, pp. 531–532). It is interesting to speculate whether *C. wattiorum* could have once been gathered from the wild as a kind of "saffron" in the same way that other wild crocuses e.g. *C. longiflorus* in Italy (Casoria et al., 1996) are used.

8 We cannot be sure, but "Actolia" is probably a miss-print of Aetolia, a region of Greece on the north coast of the Gulf of Corinth.

9 Cyrenaic: from Cyrenaica, the eastern part of Libya near the ancient city of Cyrene. Centorbi is now called Centuripe.

Now, our eastern [saffron], brought from Hungary and Austria, being of exceptionally good quality, by far surpasses all the others in excellence, see Coler, book VI of *Gartenbau*, ch. 71, p. 223; Minderer, *Aloedarium*, ch. 4, p. 19, et al.

CHAPTER VI

Properties of *Saffron*

In the past, the Babylonians had this custom, that, when they were failed by doctors, they would take the sick to public places, and expose them to the judgement of everybody so that, if by chance anyone had suffered from a similar disease, they could make known publicly what had benefited them.

Nowadays, a wretched man is for this reason exposed in the public audience-room of the University; and as Seneca, lustre of Spain, writes in *De consolatione ad Marciam*: a weak and frail body, naked, defenceless by nature, in need of the help of others, exposed to every abuse of fate, although he exercises his arms well, he is fodder for any beast, victim of anyone; composed of feeble and soft bones, polished in the exterior features, intolerant to cold, heath and toil, even with inactivity and rest he is going towards decay. In this, from beginning to end, he does not exist if not because of his habits.

Since one is worn out carrying the many and diverse afflictions of life, a secret, or a panacea, would be necessary to prolong the period of human life, which is, alas, too short (not unjustly, the dying philosopher Theophrastus blamed nature for giving a longer life to deer and crows, who do not mind at all, but a very short one to men, who mind in the highest degree).

Praise of *saffron* and its virtues

But who does not know that *saffron* has always appeared as extremely beneficial for almost all remedies against diseases? Rightfully, *saffron* claims for itself the name *Regis Vegetabilium* ["King of Plants"], and at the same time it deserves to be called also 'Enlivening Panacea.' Schenk in person, in his *Observationes medicae, De venenis ex plantis*, book VII, p. 879, confirms this. *Saffron* is a restorative medicine, which deservedly must be mixed with drugs both for internal and external use; in fact it fortifies and enriches.

That most scrupulous investigator of medical matters, Dioscorides, book I, ch. 16, p. 18, as well as Mattioli, book I, ch. 16, p. 18, classify *saffron* as warm in the second degree, and dry in the first degree. It is moderately astringent, and with this ability fortifies the heart and other parts, and with its agreeable warmth it heats, helps the digestion, and by slightly drying it benefits consumptive dispositions, and cures putridity.

Lobel, *Stirpium observationes*, p. 68, following Galen and Paul of Aegina, classifies *saffron* as warm in the second degree and dry in the first degree, and ascribes to it the merit of being astringent and helping the digestion.

 | DOI:10.1163/9789004435292_010

In the famous ancient writer Serapion, it is defined as warm in the first degree and dry in the second degree, beneficial for those affected with coughs and pleurisy.

Camerarius, in his *Hortus medicus*, claims that *saffron* is very good for conditions of the chest: quickly drinking about ½ scruple and also 1½ scruples with 1 grain of musk and excellent wine benefits the asthmatic. In fact, *saffron* is the life of the lungs: the famous Cardano, *Consilium de respirationis difficultate*, only with *saffron* set free from the jaws of death many who appeared to be destined to die within few hours due to a breathing difficulty.

For people with consumption in a critical condition, and about to die, there is a very effective remedy, which restores the breath and extends life for several days: ½ scruple, or about 1 scruple, of *saffron* taken with raisin-wine, or sweet wine. Dodoens, Avicenna, and the famous botanist J. Bauhin in *Historia plantarum universalis*, p. 639.

The magnificent and excellent Dr Paul de Sorbait, esteemed Dean of the Medical Faculty in Vienna, and Court-Physician of Her Majesty Empress Eleonora, claims that he himself often tried this remedy.[1]

In Pietro de' Crescenzi, *De agricultura*, book VI, p. 204, *saffron* is described proportioned in its qualities, as warm and dry in the first degree; although some say that it is warm in the third degree, see Ruel, *De natura stirpium*, book II, ch. 11, p. 295.

In Isaac Israeli, *De dieta et herbis particularibus*, p. 481, as well as in Simon Sethus, *saffron* is good for the stomach, and causes the digestion of food, removes obstructions, benefits phlegmatic and lethargic defects, and is useful against sores.

In Bock, *saffron* cleanses the blood, resists poison, for which reason, not without cause, it was employed against the plague; furthermore, it fights paralytic limbs and hardened nerves.

In Mattioli, ch. 16, p. 18, in cases of apoplexy, drowsiness and aphonia, *saffron* mixed with extremely sharp vinegar and castoreum, in which a feather is dipped and inserted into the nostrils, awakens from deep sleep.

According to the famous Werner Rolfinck, *Methodus medicinae*, in particular book II, ch. 5, p. 1, *saffron* softens, ripens, stimulates urine, procures an agreeable complexion, and excels in many other qualities.

According to Sennert in *Epitomes physicae*, book VI, *De historia plantarum*, ch. 6, section 6, p. 65, *saffron* removes obstructions and jaundice,

1 Empress Eleonora, second wife of Holy Roman Emperor Ferdinand II, lived 1598–1655, becoming Empress in 1627. Paul de Sorbait must have been well known to Hertodt: Sorbait wrote an approbation in Hertodt's *Tartaro-Mastix Moraviae*.

induces menstruation, stimulates urine, arouses sexual desire, strengthens the heart, brings joy, and enlivens the senses.

Additionally, it removes inflammations gone past the initial stage, for instance those of the chest, of the bladder, of the kidneys, of the liver and so on, and most particularly those of the eyes, after being mixed with an egg. It also stops lachrymation if applied with breast milk. Furthermore, it cures suffocation of the womb, exulceration of the stomach, and abscesses of the ears. It is added to pessaries and poultices for the womb and the anus. It soothes erysipelas, it helps the digestion, and it is most sought after for dyeing sauces with a delightful colour. Johann Bauhin, *Historia plantarum universalis*, book XIX, p. 639. Johannes Wirbunk in his *Artzney-buch*, p. 222, recommends *saffron* for breasts hardened by coagulated milk.

Lauremberg recommends it for the bites of poisonous animals.

Conrad Khunrath of Leipzig in his *Medullae destillatoriae*, p. 337, most eagerly recommends *saffron* for melancholia; dizziness; distress; delirium; strengthening of memory; wry mouth; to cause catarrhal fluxes to descend into the ears, teeth and the throat; for colic; hardness of the spleen; haemorrhoids; for dysentery with oil of juniper and similar; for earache with breast milk; to dry sores of the mouth with oil of myrrh; to reduce a fever with oil of violets or mixed with another corresponding ointment and smeared on the backbone; for the itch with oil of roses and egg yolk; for pain caused by foot-gout, hip-gout, hand-gout with oil of roses and opium (right where the heat is perceived).

Saffron is also an anodyne.

It is especially good in abortions, and in inducing menstruation: in fact, it has been noted by some physician-researchers, like Amato Lusitano, and others, that it tinged even the foetus in the womb, also in Simon Paulli's *Quadripartitum botanicum*, p. 52, [section entitled] "On its power in our investigations on the termination of several things."

However, *saffron*'s excessive use is hurtful: it causes pallor, food aversions, dim vision, and, in a wonderful way, it deprives men of life with laughing. Naturally, because the spirits are augmented and the skin pores opened, the inner spirits impetuously go out together with the spirits received, and are dispersed. This is caused, if I am not wrong about what he said, by a greater enlargement of the heart: in fact, the heart is pleased by the warm exhalations of *saffron*. Borel, *Observationes*, *centuria* 3, p. 303. Or, as Hoefer puts it in his *Hercules medicus*, book II, *De affectibus thoracis*, p. 83, the spirits suffocate men with their abundance.

Observations of this kind are also found in Camerarius, *Memorabilium*, *centuria* 3, § 44; Zacuth of Lisbon, *Praxis medica admiranda*, book III, observation 139; Schenck in *Observationes medicae*, book VII, *De venenis ex plantis*, p. 879, quotes from Amato Lusitano.

Thus, *saffron* causes laughing, and it should not be difficult to believe that men die of immoderate laughing because of the excessive *saffron* eaten.

In Medina del Campo, the most famous market-town of all Spain, we saw a merchant who purchased several packages or bags of *saffron* to bring to Portugal. He put lots of it in a jar containing juicy meat for dinner, bursting into such a violent laughter after eating it, that, at the time, he almost died of laughter.[2]

The most esteemed Dr Johann Michael often said that some fell into illnesses and intoxication from excessive inhalation and usage of *saffron*. Borel, *Observationes, centuria* 3, p. 303, reports that *saffron* fumes were deadly to a servant of a merchant, who lay down and slept next to a great quantity of *saffron*. In fact, because he thought that no harm could come from it he always remained in the same workshop; from this he got such a headache, and such a weakness of heart that he passed away.

Costeo writes similar stories, that many who had used a small bag of *saffron* as a cushion died with incredible illness after developing a very bad headache.

Coler, *Gartenbau*, book VI, ch. 71, p. 223, saw many women in labour die for being exposed to an excessive dose of *saffron*. Borel, *Observationes, centuria* 3, p. 303, also heard that horses die of bloody urine when they carry large quantities of *saffron*.

---

2 The medical term for this affliction is cachinnation. The aroma of saffron was said to have a similar effect. Hempel (1865, p. 239), quoting Schulz, writes, "children were assailed with fits of immoderate mirthfulness in consequence of smelling at an empty vessel which had contained essence of crocus [saffron]".

CHAPTER VII

Uses of Preparations of *Saffron*

In Rome *saffron* is very expensive; it is the eye of flowers, the gem of plants, the glory of meadows, the ornament of the earth, and is called by some the Philosophers' spice par excellence, because of the many and great benefits that it brings to the advantage of humankind. In fact, it has always been extremely suitable for helping the digestion, for strengthening and enlivening the internal organs, and for making food enjoyable with its sweetness and colour. And to make it even more culinary and useful for seasoning food, Felix Platter, *Praxis medica*, p. 446, suggests the following recipes.

Take of: Ginger, 4 ounces
grains of paradise, ½ ounce
pepper, 2 drachms
nutmegs, cloves, of each ½ ounce
galangal, 2 drachms
cinnamon, 2 ounces
saffron, 6 drachms.

Mix.

You can make it sweeter in this way:

Take of: Cinnamon, 4 ounces
ginger, 1½ ounces
cloves, 1 ounce
pepper, ½ ounce
saffron, 2 drachms.

Mix. Make a powder.

In pharmaceuticals, *saffron* has various applications. The powder is prepared according to pharmaceutical Art with the *threads of saffron* and it is indispensable in the preparation of several drugs. Powder

If *saffron* powder is combined with the saliva of a fasting man (keeping clear from any other solution) and the area affected by erysipelas, right where pain and redness are, is smeared with it, it does wonders if a fumigation, as described below, follows.

Water distilled with *saffron-flowers*, with or without threads, is extremely useful for preventing insomnia. Compare Mattioli, book 1, ch. 16, p. 18. Dose: 4 ounces of this. Water

 | DOI:10.1163/9789004435292_011

Spirit To make *spirit of saffron*:
Take of: *Threads of saffron*, as much as you will.
Cut up and pour into rectified spirit of wine. Let them macerate for a while, then press them. Purify the expressed liquid through digestion and sedimentation, and decant what is pure; distil the spirit impregnated with tincture.

And although there are many pharmacists who boldly deny that any of the properties of *saffron* are preserved in the distillation, I nevertheless testify, having experienced the contrary, that they talk nonsense: in fact, the *spirit of saffron* is no less efficacious than *saffron* itself when it is administered in solid form and it retains the flavour, the odour and the properties of *saffron*, Frans Dekkers in *Notae ad praxim Barbettianam*, p. 210.[1]

Tincture/ Extract The liquid that is left at the bottom from the distillation is called liquid essence or tincture.

This is more often called extract. Hartmann, *Practica chimiatrica*, p. 25; Grüling, *Florilegium*, p. 273; Mylius, *Antidotarium medico-chymicum*, p. 70; Malachias Geiger, *Microcosmus hypochondriacus*, p. 39.

This well-known *essence* or *extract of saffron*, with its extraordinary power, strengthens the heart, induces sleep and soothes the cough. In cases of pleurisy, heartburn, melancholy, jaundice, erysipelas, retention of the menses, abdominal pain and colic, it is the best remedy to get better, moreover it arouses sexual desire, and chases away sadness wonderfully. Dose: from 1 to 6 grains in appropriate vehicles. Daniel Angelus, *Catalogus medicamentorum spagyricorum*, p. 70 § 6.

Others extract the *essence of saffron* with brandy and they declare it the best medicine against melancholy, when somebody reaches a point of despair. Croll and Friedrich Hoffmann, *Methodus medendi*, p. 200.

Others extract the *essence of saffron* with plain water, rose-water, or orange-flower-water, and clarify the mucilage left in the infusion (which with time can make it rancid) with egg white, and thicken the essence until it reaches the consistency of honey. And Angelo Sala, *De essentiis vegetabilibus*, section 2, ch. 6, p. 20, considers this preferable to the essence of *saffron* made with spirit of wine.

Zwelfer in his *Pharmacopoeia regia*, p. 232, suggests another way of producing *essence* or *extract of saffron*, which carries fully the virtues of *saffron*:

1 Hertodt refers to Frans Dekkers, though this should be Frederick Dekkers.

Take of: *Austrian saffron*, 3 ounces.
Put it into a glass alembic and pour over: rectified spirit of wine and alcoholic spirit extracted twice in a sufficient quantity from fresh flowers of lily of the valley, orange-flower-water distilled not using a steam-bath, of each 1 pound. Let them macerate for a few days on gentle heat and then remove all the spirit added and the water by distilling in a water-bath or in a steam-bath; keep this. In the meantime, extract the *saffron* left at the bottom of the alembic with a sufficient quantity of spirit of wine poured into the alembic repeatedly, until all the *tincture of saffron* has been extracted. Afterwards, filter all the extractions or coloured spirits in the glass alembic and with a water-bath bring them to the consistency of an extract. Now dissolve this extract with the spirituous water extracted from *saffron*, filter it and store carefully this *essence of saffron*, which is as excellent for cold dispositions as it is beneficial for warm dispositions.

Alternatively make the extract in the following way:
Take of: *Threads of saffron*, 2 ounces.
Cut them up, put them in a glass cucurbit, and pour over them hot spirit of black cherries, 1 ounce. Let it macerate a day and a night, then powerfully distil all the spirit in a water-bath. Add spirit of wine or cordial water to the *saffron* left in the alembic, repeatedly, till when the spirit or the water becomes coloured. Extract the essence, which, filtered in the gentle heat of the bath, is to be brought to the consistency of a very liquid syrup. Pour the residue into a porcelain vessel or a glass diploma and let it evaporate further until it reaches the compact consistency of *extract of saffron*. Finally, add the spirit of cherries previously extracted and kept, and make the extract of the right consistency with *oleosaccharum of saffron*, 2 drachms. Dose: several grains in an appropriate vehicle. In addition to the virtues already mentioned above, you will find this extract extremely serviceable in cases of disorders of the womb and conditions of the brain.

Method of preparation of the *arcanum*, or *quintessence of saffron*. Arcanum

In the autumn, when *saffron* flowers, take of: Threads, a good quantity, pour spirit of wine, free from any phlegm, to the height of 4 digits, into a glass vial carefully sealed for one month. Afterwards distil the extremely fragrant ethereal spirit with a very gentle fire of ashes until there are no *striae*.[2]

2 *Striae* translates as furrows, pleats or stripes. It is unclear what is being meant here, but perhaps this is a reference to the formation of "legs" (as seen when swirling wine around in a glass), indicative of the presence of alcohol.

Pour this, with the remaining damp dregs, once again into a perfectly closed vial, and let them digest in lukewarm ashes or in a lukewarm bath for 4 or 5 days, and distil again until, as said above, there are no *striae*. Be careful not to distil to such a point of dryness that the *saffron* is burned up and the empyreuma smell comes, and if a bit of phlegm comes up, dispose of it by pouring the abovementioned distilled spirit in again, which again after the second digestion is distilled without phlegm, and repeat a fourth time the affusion, digestion, and distillation of the spirit, and thus, with the ethereal spirit, the *volatile salt of saffron* will rise. Both of them need to be kept united and well closed.

Volatile salt of saffron

Now, take of: A good quantity of *Austrian saffron* dried in the shade or with gentle heat; distil oil by placing it into an earthenware or copper pot appropriate for descending distillation, and, according to Art, far away from extremely strong fire, until it is consumed. Separate the oil from the fetid water stinking of empyreuma by means of a funnel and further purify it with a little bit of vitriol calcined to redness by means of a retort, the vitriol being transformed three or four times in single rectifications, until the oil gets rid of the empyreuma. Put into the oil new and fresh *threads of saffron*; after they have stayed in the sun for several days, press, put in new threads, let it stay in the sun, and press, and repeat this three or four times, until the oil acquires a strong odour of *saffron*. Then join this very oil to ethereal spirit impregnated with *volatile salt of saffron*, and finally add fixed salt from the ashes remaining after the distillation of the oil (take this in a small quantity), free from all recrements after several repeated dissolutions, calcinations, filtrations, and evaporations, white, and pure. Keep the *arcanum*, or *quintessence of saffron*, in perfectly closed vials for a while.

Fixed salt

Enlivening homunculus

Johannes Tilemann in *Appendix de materia medica*, p. 300, describes the following enlivening homunculus of *saffron*.

Take of: *Best saffron*, 8 ounces.

Put it, finely chopped, in a glass cucurbit and pour over filtered milk whey, or a suitable spirit, or lemon juice, distilled, to the height of 2 or 3 digits; leave it for 12 days and nights, or longer. Mix the tinctures, extract the salt from what is left, put together with the tinctures in a closed vial, let it coagulate to powder, or not. Dose: up to ½ scruple.

Moreover, *saffron* improves many mixtures to such a degree that you will find very few of them by Mynsicht and Zwelfer in which it is not included.

Before anything else, let us attain an elixir created by Paracelsus, known to everybody, which the author prepares from *saffron*, myrrh and aloes, adding spirit of wine and spirit of sulphur, and is good against all diseases, both warm and cold. It strengthens internal and external senses, invigorates the stomach and the intestines, cures the afflictions of the head and catarrh, relieves feverish subjects and cheers the body.

Helmont observes this fault in the preparation, namely that this remedy was not effective on account of the quantity of tree of life with which only it could promise the hope of a long life, renovations and rejuvenation.[3]

But he himself provides its secret preparation as follows:

Take of: Bright aloes, best myrrh
best saffron, of each 1 ounce.

In fact, if you have taken more, you will find that you have done it in vain. Rub the first two accurately, but as for the *saffron*, because it cannot be rubbed if not dried, press it together into a ball by pounding. Put them in a very large and sturdy glass, and seal it by melting the neck of the glass. Distil it with a moderate heat, so that the vessel does not break, until you see that all the mass coalesced at the bottom, and clear oil forms a circle with water on the sides of the glass. Then open the neck of the glass and pour in cinnamon-water, 1 pound, and distil by means of wet sand, over which boiling water is to be poured gradually, until nothing more drips from the beak of the alembic.

And with this remedy both quartan and continuous fevers are immediately broken, to such a degree that someone who at night had been administered the Holy Viaticum and the anointing of the sick, had me as guest for lunch near the bed. This says Helmont.

The Reverend Fathers of the Order of Minims of St Francis of Paola in Vranov, a place near Brno scattered with stone, who toiled greatly till the present in the study of the secrets of nature by means of fire, strongly believed that Paracelsus' authentic preparation is the following:

Take of: Our menstruum, as much as suffices
myrrh, hepatic aloes,
saffron, of each 1 ounce or more.

Let them digest for three months in a mortar well shut, filter them, and expel what is left by means of a retort.

3 Helmont (1579–1644) concluded, via his alchemical investigations and finally as the result of a dream, that the tree of life was the cedar of Lebanon (*Cedrus libani* A. Rich) (Hedesan, 2016, p. 189 et seq.).

Our Menstruum

It is made of vitriol of mercury, whose sulphur is extracted with a long digestion with spirit of wine, by digesting and expelling, by means of a retort, until it assumes a fragrant odour. This usually happens after the sixth time. So far we have made it with spirit of salt and we have tested its efficacy against various conditions, especially against dispositions of the stomach.

Encomium of salt

Indeed salt is the soul, as Deodatus writes in *Pantheum hygiasticum*, book I, ch. 18, as much as the balsam of nature, which is superior to any other thing in virtue, it becomes miraculous by necessity, and it is made worthy of veneration for its pleasing innocence. And if you are testing virtue? Salt is the fiercest enemy of any corruption, healthy, a renewed remedy to fragility, and promoter of fertility. Its virtues should be admired rather than counted: as it boasts as many marvellous things as the grains it is given in numerous abundance.

Nature granted that salt should by all means be given to the majority of things, and that it should be indispensable to all. It is overflowing in the immense cavities of the sea; it is abundant in the mountains; plentiful in springs and everywhere perpetual. And the frame of the world would inevitably collapse if it was not strengthened by salt. Even the rivers, which, straying about in isolated places with their continuous and impetuous course, full of waves, measure distances immense in size almost in secret, and rushing with effort they flow into the sea not on account of motion, so that they are not corrupted by dishonourable idleness, but so that they are cured by the admixture of salt into them, because they contracted due to the immoderate presence of sweetness.

Indeed, salt is healthy! This is proved even by the Scriptures: the Church, which cured diseases with the admixture of salty water, and, following an uninterrupted tradition, proceeds to the sanctification of the water with the virtue of our friend the salt, knows this. Then, it is a fact that waters are maintained unspoiled by the association with salt, otherwise when these have been spoiled, the earth would become dirty with infected air, and, dirty, reduced to ashes, it would deny its fruit in a worthless torpor, it would take away the food, and, finally, when the general begetting of all things has stopped, our sub-lunar particle, very close to ruin, would perish. And if you are testing its innocence? You will find that it alone is blessed: in fact, while fire consumes the things that it meets with wicked voracity; air destroys all things with an idle battle; water devastates the other things with impetuous efforts; the earth turns well-ordered things upside down from the foundations with unexpected movements; gems fitted with poisons maliciously trouble the stomach with pains, and the head with anguishes; gold spoils the entire world with pleasing wounds; only the innocence of salt does not know how to hurt, even if it wanted to.

Therefore, deservedly, both because of its extraordinary virtue and the unparalleled need of it, and because of its pure innocence, it should be regarded as of value before all things. We trample upon the earth with our feet, we make water filthy with impurities, we impetuously force the air in and out, we spit on the fire without fear, bread is thrown to the dogs not given. The extremely powerful lineage of the Turks dreads wine. Gems, gold and everything which is valued by the senselessness of few is held in slight esteem by the good sense of many. But only the excellence of salt neither receives nor suffers dishonour, in fact salt is on earth what the sun is in the sky, and as Hermes Trismegistus said,

> *"In sole & sale natura sunt omnia"*
>
> ["In nature all things depend upon sun [fire] and salt"].

But let us go back to *saffron*, which did not want to be absent in Croll's elixir for the womb; the author recommends this very elixir as a remedy for various conditions and suffocations of the womb, and defects of the menses.

And Mynsicht did not neglect *saffron* in his elixir of sulphur, which does wonders to conditions of the chest and of the lungs. Namely: it cleans the chest from impurities and catarrh; it cures coughs, both old and recent; it helps those affected by consumption, excessive melancholia, sadness, heartburn and suchlike; it brings joy; it is effective with a long-lasting coarctation caused by difficulty of breathing, and with chronic asthma. Additionally, it strengthens the heart, protects from putridity, prolongs life and delays old age. *Armamenta chymica*, p. 271.

Johann Michael, in the famous *Medicinae oraculum*, as well, includes *saffron* in his elixir for the chest: Michael's elixir for the chest

Take of: Lungwort,
mountain cudweed, horehound,
white horehound, sundew,
scabious, leaves of coltsfoot,
tops of scabious, of each 1 handful

flowers of elecampane, scabious, of each 3 pugils; roots of elecampane, coltsfoot, round birthwort, orris, of each 1 ounce; choice red myrrh, choice mastic, *eastern saffron*, liquorice juice, of each 1 ounce; cinnamon, choice cardamom, of each ½ ounce; best gum Benjamin, 1½ ounces; storax, ½ ounce; alyptae moschatae, 2 drachms; nettle seeds, 3 drachms; cress seeds, 2 drachms; sugar candy, 4 ounces. Besprinkle them with spirit of sulphur, pour over spirit of hyssop or of elecampane. Make an elixir, according to Art.

To these remedies we are going to add *oil of saffron*: cooked, insolated, and distilled.

Cooked *oil of saffron*

The physicians of Augsburg, p. 318, say that it should be cooked as described by Mesue:

Take of: *Saffron*, ½ ounce
choice myrrh, 2 drachms

vinegar, as much as suffices, to be soaked for 5 days; cardamom, 3½ drachms; oil, 8 ounces. Let the *saffron* and myrrh soak for 5 days, on the sixth day add cardamom, on the seventh day cook with the oil on gentle fire until the vinegar is consumed. This very oil strengthens the nerves and the womb, and soothes their pain, is useful for convulsions and injuries of the womb, dissipates the absence of feelings, and also preserves a vivifying complexion in the body.

Another cooked *oil of saffron* for oedematous feet

Another *oil of saffron* for oedematous and dropsical tumours of the feet, used very often, is the following.

Take of: The simple oil of scorpions, 2 ounces
waters of mugwort, houseleek, of each 1 ounce
saffron sufficiently powdered, ½ drachm.

Cook by turns until the water is consumed and the oil has been dyed well by the *saffron* and impregnated, taking care that the *saffron* does not burn. Use it for the anointing of the feet, especially the soles of the feet.

Oil of saffron by insolation

Oil of saffron by insolation

Take of: *Saffron threads*, as much as you please.

Then press them into a closed glass with oil of sweet almonds, expose the glass to the sun and, after placing a mirror in the back, let it boil out from the rays reflecting back from the mirror, avoiding the rays of the moon at night. After three days, press very well, and let the oil be impregnated with new threads and proceed as described above. Repeat the process a third time with new threads, and then use it as an extraordinary painkiller.

In addition to the method mentioned above under *arcanum of saffron, distilled oil of saffron* can be prepared in the following way: *Distilled oil of saffron*

Take of: *Saffron*, as much as you please.

Besprinkle it with aqua vitae, let it soak. Then distil over a slow fire, and, the distilled spirit having been by now completely removed, when the oil has started to distil, put the *saffron* taken out of the cucurbit in a small bag, carefully bound by threads, and press it with a suitable press, after having heated well the two iron plates of the press. Distil, rectify and let circulate until pure oil is separated from the thicker substance. Then the dregs can be digested again or caused to putrefy with the distilled spirit previously separated, and distilled again. And this method is considered extremely convenient and useful by Ryff, *Thesaurus euonymi*, p. 317.[4]

Others proceed in this way:

Take of: *Austrian saffron*, 1 pound.

Let it dry and slightly powder it, add tartar and salt, of each ½ ounce; after moistening them with egg white, mix until they become like a sauce. Put them in a cucurbit, pour over them lukewarm water, or rather honey-water (by which they say that oil is made extremely fair like the sun), close well, let it digest over a slow heat for three days and nights. Then open, and, with the alembic bedaubed from above with sand, distil gradually, and thus a red oil will come out. Conrad Khunrath, *Medullae distillatoriae*, p. 333.

Philip Jakob Sachs von Löwenheim, our Phosphorus and glory of the *Curiosi*, on December 20th, 1669, shared with me in a letter another method to prepare *distilled oil of saffron*. This is his description:

Take of: *Eastern saffron*, ½ pound.

With egg white make it like a pottage; add salt of tartar and burnt crystal salt, of each ½ ounce. Mix in salty water released from butter, **das es nicht zusammen gehet** ["this does not go together"]. Digest in a bain-marie, then add: spirit of wine, 2 ounces. Distil over a slow fire.

Those who wish to know the properties of this oil, refer to the previous chapter.[5]

From rectified *oil of saffron*, 2 drachms, or as much as suffices; white sugar, 2 ounces; if you pour the oil by drops continuously so that it can stay in *Oleosaccharum of saffron*

---

4 *Thesaurus euonymi* was not written by Ryff, but authored by Conrad Gessner.

5 Strother, 1727, p. 169, writes that the distilled oil of saffron was little used (presumably due to its expense) because it needed 10 pounds of saffron to yield just 1 scruple of oil.

the form of powder, you make oleosaccharum, extremely effective against lung diseases.

From *oil of saffron*, white Canary sugar and starch, as much as suffices, you make a very small confection, called *confectio sine anima* ["confection without spirit"], of the same power as the oleosaccharum.

Magistery of saffron

According to pharmaceutical Art, *magistery of saffron* can be prepared in the following way.

Take of: Strong, well clarified, lye prepared from salt of tartar with rain water.

Pour *saffron threads*, dried in the shade and cut up rather coarsely, as much as you please, into a glass; let them stay in a tepid place, until a tincture is satisfactorily extracted. Precipitate the lye when it is well filtered and still warm with the affusion of warm rain water, in which raw alum has been dissolved. Then, sweeten very well and keep for use.

Wine of Saffron

The following *wine of saffron* is the miraculous remedy in cases of jaundice:

Take of: Malmsey, 1 sextarius
egg yolks, 2 in number
Austrian saffron, 1 drachm.

Mix everything together; take half of the dose when going to bed and the rest in the morning.[6]

Vinegar of saffron

Those who wish to make *vinegar of saffron*:

Take of: *Austrian saffron*, 1 ounce.

Cut up, put in a glass cucurbit, pour over best wine vinegar, 2 pounds. Let it digest in a warm place for one or two days, then, with a water-bath, distil by placing 1 drachm *Austrian saffron* onto the neck of the alembic.

Oxymel of saffron

This excellent vinegar is exceptionally suitable for the preparation of pectoral oxymel or oxysaccharum, if an equivalent quantity of pure honey is added to it and it is boiled in a glass vessel by simple composition until it reaches the right consistency of the oxymel.

Hydromel of saffron

Hydromel of saffron can be thus prepared:

6 Wine of saffron was later replaced in pharmacy with tincture of saffron, made with brandy. This was because the acidity of the wine would eventually decolourise the saffron, whereas the tincture would keep its colour for longer (Lewis, 1778, pp. 294–295 & 309).

Take of: Spring water, 24 pounds or 32
honey, 2 pounds. Mix.

Boil over gentle heat removing the foam with a spoon until the third or fourth part of it is consumed, and until a fresh egg put in does not sink but floats without doubt. After boiling add:

saffron bound in a nodulus, ½ ounce

Then filter through a Hippocratic sleeve in order to remove the thick dregs, pour into a large jar, and leave to ferment and depurate in a warm place or behind a furnace for four or six weeks, then store it in a store-room. It is good for phlegmatic and asthmatic persons.

Syrup of saffron

The *syrup of saffron-flowers* is extremely good for both wet and dry dropsy, as well as for diseases of the spleen and erysipelas.

Take of: Fresh *flowers of saffron without threads*, as much as you please.[7]

Pound in a marble mortar, quickly pouring waters of mugwort, flowers of peony, acacia, balm, of each as much as suffices, so that they reach above the flowers. Leave them in a closed vessel for 48 hours stirring twice a day, then press out the remaining flowers and add just as many new and fresh flowers. Make an infusion and a colature and repeat this a second and a third time. Then take this colature and with clarified sugar, as much as suffices, make a syrup of the right consistency.

We will not be against those who should wish to prepare this syrup following the method of the most skilful Zwelfer as suggested in *Pharmacopoeia Augustana*.

We have seen that this very syrup prepared according to the procedure mentioned above has been used with great success against the aforesaid diseases.

Eclegme of saffron

Mynsicht, *Armamenta medico-chymica*, §15, p. 252, considers the *eclegme*, or *lohoch of saffron*, to be extremely good for warm, irritating, delicate and acute rheum from head to chest, and that descending the trachea, for coughs, dryness, for inducing sleep, for mitigating the vehemence of the heat, for strengthening the spirits, and for preserving inborn warmth, and extremely beneficial in cases of most acute pain, especially colic, nephritic, pleuritic, arthritic pain, in cases of excretion of blood and excessively abundant menses, and also with burning fevers, diarrhoea, as well as with dysentery caused by an irritating and bilious substance. Dose: from 1 drachm to 2 drachms, 3½ ounces.

7 These flowers are the by-product of removing the threads.

Electuary of saffron

There are several descriptions of electuaries made from *saffron*, to such a degree that (as Jean de Renou, book III, *Commentarius* p. 211, says) the pharmacist who does not pride himself with having the best description of it is not considered ignorant.

It is described in Johannes Bockel's treatise *De peste*, ch. 21, p. 185; Schwenckfeld, *Theriotropheum Silesiae*, p. 273; Tabernaemontanus; Sennert, *De febribus*, p. 640; Michael Döring, note to Willich, *De peste*, p. 171 et seq.; Platter, *Praxis medica*, p. 185; Duchesne, *Pharmacopoeia restituta*, ch. 22.

Jean de Renou, book III, ch. 10, p. 210, has the following:

Take of: *Saffron*, 1½ drachms or 2 drachms.

Insert it with egg yolk in an eggshell, only open in the one part from which the albumen has been taken away, and afterwards closed with another eggshell or with paste, then roast it in the oven. Once the substance has been removed from the shell, reduce it to powder. To this add dittany, tormentil, of each 2 drachms; hartshorn, nux vomica, of each 1 drachm; angelica, zedoary, grains of juniper, elecampane, of each 2½ drachms; cinnamon, mace, of each 1½ drachms; camphor, 1 drachm; treacle, 3 drachms. Mix. Pound everything vigorously, and add syrup of lemons, as much as suffices, make an electuary, very much recommended against plague.

I add another *electuary of saffron*, frequently used against cold diseases of the chest.

Take of: Condited cassia fistula, 1½ ounces
magistery of saffron, ½ scruple
oleosaccharum of the same, 6 grains.

With *hydromel of saffron*, as much as suffices, make an electuary.

Conserve of flowers of saffron

The *conserve of saffron-flowers* is prepared with flowers cut into tiny pieces with the addition of the finest sugar, twice the weight of the flowers. Expose to the sun in a sealed jar, and shake from time to time.

Pills of saffron

Saffron is one of the ingredients of the very much praised Emanuel's pills, which strengthen memory, enliven the vital spirits, are very effective against colic and plague, etc. This is the recipe:

Take of: Hepatic aloes, 1 drachm
red myrrh, 2 grains
Austrian saffron, 1 grain
mithridate of Damocrates, 2 grains.

With potent wine make a mass. Shape the pills like white pepper-corns. Dose: 12, 14 or 15.

Wilhelm Fabry, *Observationes chirurgicae, centuria* 6, p. 530, describes the following pills for invigorating the noble body parts.

Take of: Finest Socotrine aloes, 3 drachms
choice rhubarb, 5 scruples
myrrh, *saffron*, of each 2½ scruples
cream of tartar, ½ drachm; confection of kermes-berries, 1 scruple
oil of anise, cinnamon, of each ½ scruple.

Mix. Make a mass, according to Art, with as much as suffices of extract of water-germander.

The properties of these pills are diverse and manifold: namely, they invigorate the various noble parts of the body extraordinarily, they gently cleanse the bowels, they draw the thick and sticky phlegm out of the head and of the other parts, they fight putridity, and are comparable to a balsam with which we can stop and shatter the blows of many diseases.

Troches of saffron

The physicians of Augsburg list Nicolai's *troches of saffron*, p. 202.

These troches remove obstructions in the spleen and in the liver, dissipate bloating, prevent the festering of humours, and strengthen the heart. The following moschardini are the greatest arcanum against asthma:

Moschardini of saffron

Take of: *Saffron*, 1 scruple
musk, ½ drachm.

With as much as suffices of sugar and gum tragacanth dissolved in a water-decoction of root of iris, make troches or moschardini.

Round tablets [rotuli] of saffron

For *round tablets of saffron*:

Take of: *simple extract of saffron*, 1 scruple
distilled oil of the same, 5 drops

finest sugar dissolved in a decoction of roots of orris from clarified coltsfoot-water; make round tablets with or without musk.

Tablets [morsuli] of saffron

For *tablets of saffron*:

Take of: Condited elecampane roots, ½ ounce
condited orange-rind, 2 drachms.

Cut up, boil both in pectoral water, as much as suffices. Dissolve finest sugar, ½ pound, in the strained decoction, and boil again to a hard consistency. After removing from the fire, add condited orange-rind cut in tiny pieces, ½ ounce; *arcanum* or *quintessence of saffron*, 1 scruple; make tablets with or without musk.

Enema of saffron

For enemas:
Take of: Rainwater from around the time of the summer solstice, 1 pound.
Insert on its own when the bowels need only stimulating, but when, in cases of colic, the pain needs calming, make a clyster with this water tinged in some degree with *saffron*.

Where does the laxative property of rainwater collected around the time of the summer solstice come from?

Are you wondering why rainwater collected around the time of the summer solstice has laxative properties? I believe that this property must be attributed to the salty and nitrous mercurial produced by the moist vapour, which is particularly active at that time, having been exposed to the sun for a long time, but is, instead, very much dispersed by the warmer season that follows.

Croco-magma

There are several descriptions of the *ointment of saffron*, as well as of the crocomagma, among which Bauhin, *Historia plantarum universalis*, p. 640; Minderer, *Aloedarium marocostinum*, pp. 23–24, et seq.

Crocomagma has an amazing scent; it smells slightly like myrrh, it is heavy, dark, free from woodiness, slightly bitter. When soaked it gives up the colour of *saffron*, and when tasted it tinges teeth and tongue very much. That imported from Syria lasts several years.

Among its properties it has the virtue of removing cloudy vision, it induces urination, it is warming, it promotes digestion, it softens, to a certain extent it manifests the properties of *saffron* in proportion, since crocomagma claims the largest part of *saffron* for itself. Dioscorides, *De materia medica*, book II, p. 33.

Ointment of saffron

An ointment exceptionally commendable for diseases of the chest can also be made from an onion hollowed out and refilled with *saffron* and with frankincense, well mixed. From the excellent and most esteemed Dr Sachs von Löwenheim, personal communication.

Poleman, *De mysterio sulphuris philosophorum*, p. 609, describes another frequently used ointment, which, if prepared well, obtains great praise and is of great service to the neglected sick:
Take of: Honey well skimmed, 4 ounces
juice from leaves of greater plantain, 12 ounces
sweet vitriol of copper, 2 ounces.
Boil over a slow fire until it becomes thick like an ointment, then add:
best ground saffron, ½ ounce.
Then remove from the fire and well mix in the *saffron*, so that it is incorporated. Make an ointment.

We put the following effective *liniment of saffron* on feet affected by oedema and dropsy. *Liniment of saffron*

Take of: Fresh, succulent little *bulbs of saffron*, beaten to a pulp, 1 ounce
the simple oil of scorpions, 2 ounces
waters of mugwort, houseleek, of each 2 ounces.

Boil them one after the other until the water is consumed. Make a liniment.

The plaster of oxycroceum is included in the *Pharmacopoeia Augustana*, p. 384. It is useful for bone fractures, muscle pains, and long-standing abscesses; it is also good as a plaster for a wound and for calluses. *Plaster of saffron*

This *plaster of saffron with galbanum* should not be overlooked either. Mynsicht, *Armamenta medico-chymica*, § 34, p. 811. This is considered extremely useful and renowned for softening and weakening hard, swollen and strumous tumours, for calming pain of the shoulders, chest, breasts, belly, liver, spleen, kidneys and womb, caused either by flatulence, crude humours or dry temperament.

We move on to the *plaster of oil of saffron with frogs* described by Giovanni da Vigo.

Dr Johann Michael, a very famous practitioner in his time, claims that it is extremely effective against headaches, as well as against syphilitic tumours.

And do not be afraid that it contains mercury; in fact that drags out the harmful moistures, as the abovementioned doctor demonstrated.

You can also make a frontal with *saffron*, applied like a crown and very useful for inducing sleep. *Frontal of saffron*

The *sief of saffron* attributed to the French physician Dr Turquet is this: *Sief of saffron*

Take of: Sarcocolla nourished with milk,
gum Arabic, of each 6 drachms
Venice ceruse reduced to fine powder in a porphyry mortar with rose-water and made into pastilles, 2 ounces
saffron, purified opium, of each ½ ounce
camphor, 1 drachm.

Extract *tincture of saffron* in a sufficient quantity of rose-water, dissolve opium in it, and with these mix very finely ground gum. Reduce everything into mucilage, grind the ceruse with them in a mortar, and at the same time dry them. Add camphor to the powder, as well as a sufficient amount of seeds of psyllium and quince; extract with nightshade-water; make a sief. Dissolve in cold water for pain and inflammations of the eyes. Dose: 1 scruple per 1 ounce of water.

Fumigation of saffron

The *fumigation of saffron* is extremely effective for erysipelas, if the area affected has been previously smeared with *powder of saffron* mixed with the saliva of a fasting man. It can be prepared in the following way:

Take of: *Saffron,*
flowers of wild poppies,
frankincense, amber, of each 2 drachms
dragon's blood, ½ ounce.

Mix. Make thick powder for fumigation.

Cloth of saffron

The extent of the efficacy of Mynsicht's *cloth of saffron* in cases of gout in the feet and of erysipelas, if the cloth is placed onto the affected area previously smeared with Venice soap, premising what is to be premised, is discussed in his *Armamentarium chymico*, p. 525.

And this is all about the uses of preparations of *saffron* and its properties in general. Now it remains [for me] to go into more detail about its use, how it cures nearly all the sick parts of the human body with its superior power, starting from the diseases of the brain.

CHAPTER IIX

Diseases of the Brain

1 Cephalgia, or Headache

A still head is like the top of a mountain, which catches the very gentle breeze of the sky, the clear rays of the sun, the limpid gushing waters of springs and so on; in contrast, a head tormented by pain or disturbed in whatever way is a marketplace full of excitement, where the noise of those who come and go, the neighing of camels and donkeys, the din of silversmiths and coppersmiths beating [metal] are extensive. Chrysostom, *De mente irata et tranquilla*.

What is a headache?

A headache is indeed a disagreeable feeling, caused by a continuous relaxation in the membranous parts of the brain, by phlegm, bile, melancholy, serosity, air, plethora; or, according to modern medical writers, by the corrupted nervous fluid, by which nerves and meninges are moistened, troubling them with its sharpness, saltiness or thickness; or by an altered and corrupted spirit in the adjoining relaxed parts. In fevers, it is caused by boiling-up blood lingering its circulation (which in some measure bends sideways in the head) in the meninges and swelling them up.

On this account, when phlegm or, according to modern medical writers, very thick nervous sap, becomes harmful, the following pills of the excellent Dr Matthias Franz Hertodt, medical practitioner, are specific and proven to get rid of the cause, to strengthen the brain and to reduce the pain:

Take of: Mynsicht's aloëphanginae pills, 1 drachm
mass of pilulae cochiae of gum ammoniac,
Minderer's marocostinae

(This recipe is improved twice by *saffron*, clearly, as Mynsicht's aloëphanginae pills and Minderer's marocostinae both contain *saffron*, whose properties the author thus discuss in his *Aloedarium*, ch. 4, p. 30. *Saffron* corrects and makes even the humours still and without agitation, without inflaming anything. At the same time, it makes the spirits pure and clear, it restores weak spirits and those running out, it revives consumed spirits. From *saffron* comes appropriate help to the heart and the chest, cleansing of blood, and evenness of humours, which are good for the balance of health), of each 1 scruple.

extracts of black hellebore, catholicum,
panchymagogum, of each ½ scruple

 | DOI:10.1163/9789004435292_012

scammony, 6 grains
magistery of amber, 1 scruple.

With essence of spermaceti, make pills of the size of a small pea, to which at the end add:

oil of amber, tincture of silver, of each 3 drops.
Mix for 3 doses.

After using the latter, the forehead can be anointed with Zecchio's liniment (which also contains *saffron* among other things). Very often, with only this, he cured long-lasting or even extremely persistent internal and external pain in the head, after the digestive tract had been cleansed with the common remedies. He describes its recipe in the *Consilium tertio*.

Or

Take of: Balsam of Peru, 3 drachms
extract of saffron, ½ scruple. Mix.

Let it be marked: **Haubtschmerz stillender Balsam** ["headache soothing balsam"].

For headaches depending on bile, or blood, or according to modern medical writers even on sharp corrupted nervous sap, premising what is to be premised, it is evident with a bleeding and a milder cleansing from the outside, the following frontal which calms the pain and induces sleep when placed on the forehead, is worthy of praise:

Take of: Seeds of white poppy, henbane, of each ½ ounce
bitter almonds,
peach-kernels, of each 2 drachms

waters of lettuce, sorrel, water-lilies, of each 3 ounces; vinegar of roses, 1 ounce; opium, 1 scruple; *saffron*, ½ scruple. Mix.

Or

Take of: Rose-water, 1 ounce; flowers of elder,
fennel, of each ½ ounce
1 egg white
saffron, ½ scruple. Mix.

Apply on the forehead and on the temples. Horst, *De tuenda sanitate*, book II, ch. 1, p. 194.

Additionally, Lange's alabastrine ointment with juice of vervain and *saffron* is very good. See the hiera picra of our most esteemed President, p. 81.

Others prepare an ointment with the yolk of a fresh egg cooked under ashes, mixed with *saffron threads*, and oil of roses. They apply this to the temples with blue paper, happy with the result. Communication from the most esteemed Sebastian Georg Jung, our most pleasant fellow practitioner in Vienna.[1]

Or

Take of: Conserve of red roses,
ointment of poplar-buds, alabastrine ointment, of each 3 drachms
oil of nutmeg, ½ scruple
saffron, 1 scruple. Mix
and apply on a piece of tow.

Or

Take of: Dried flowers of elder,
water-lilies, red roses,
pot barley, of each ½ ounce
white amber, 3 drachms
saffron, 1 scruple
camphor, Thebaic opium, of each ½ scruple.
Make a fine powder, add:
oils of water-lilies,
bitter almonds, of each 6 scruples
vinegar of elder, best rose-water, as much as suffices.

Boil in the fashion of a poultice; apply to the forehead for headaches with hot flashes, delirium, wakefulness. Angelo Sala, *De peste*.

To cure and prevent headaches, women in labour bind to their foreheads frontals soaked in rose-water, *saffron* and camphor, and impregnated with them, and dried again. Rodrigo da Fonseca, *Consultationes medicae*, p. 368, recommends the following remedy against headache:

Take of: Myrrh, *saffron*, euphorbium, of each 3 grains.

Pound, mix with egg white, bind around forehead and temples.

Or

Take of: Scammony, 1 drachm
myrrh, costus, of each 2 drachms
saffron, 1 drachm

1 Hertodt has reversed the name of his colleague here; it should be Georg Sebastian Jung.

verjuice, 1½ ounces
opium, 4 grains.

Carefully make all these things into a cataplasm, with some drops of sharp vinegar and white wine and egg white. Zacuth of Lisbon, *Pharmacopoeia*, p. 85.

In desperate cases of long-standing headaches, Kyper in his *Collegium medicum* prescribes the following cataplasm, which matches expectations:

Take of: *Saffron*, opium, of each equal parts
(for example, 1 drachm), pound together
in a mortar, mix and add
egg yolks (for example, 7).

Mix these things with spirit of wine, 1 ounce; reduce to a cataplasm. Alternatively, instead of egg yolk, because it sticks together with difficulty, take melted wax and turpentine.

11 Paralysis

What is paralysis?

Paralytic people, whose rigid limbs totter and whose hands, common means of many actions, are deprived of their strength, and whose very feet, which used to support their body, are not given to be set fast, are an example of how feeble the structure of the human body is. Similarly, a fault in the foundations is the most dangerous fault in a building. In fact, howsoever you fix the other things, this brings ruin, and, ruin impending, nothing remains but leaving the collapsing abode as soon as possible. Petrarch, *De aegritudine tibiarum*, dialogue XCV. Paralysis can be of movement and sensation, or of both at the same time, or it can be a numbing in several parts. It is caused by the afflux of animal spirits into the nerves being blocked by thick, viscous, or icy humours or by humours imbued with numbing poison, or by thick vaporous substances obstructing the nerves.

In cases of paralysis, premising what is to be premised, the following balsam helps:

Take of: Bay-berries, castoreum,
eastern saffron,
euphorbium, of each 2 drachms

cloves, mace, of each ½ ounce; mastic, myrrh, liquid storax, of each 6 drachms; white amber, gum elemi, galbanum, of each ½ ounce & ½ drachm; fresh earthworms, 22 in number; cypress turpentine, ½ pound; spirit of wine, 5 pounds. Distil according to Art. Make a polychrestic balsam.

And for all the affections of the nerves and the deficiencies of vital energy, that extraordinary liquor invented by Mesue is not to be disregarded either. When the backbone and the area where the nerves originate are anointed with it, broken, contused and dying limbs are restored and nearly recover. This is its recipe:

Take of: Myrrh, hepatic aloes,
dragon's blood, frankincense,
true mummy, opopanax, bdellium,
spikenard, carpobalsam,
saffron, mastic, gum Arabic,
liquid storax, red storax, of each 2½ drachms

musk, ½ drachm; turpentine, to the equivalent weight of all of them. Grind these and put everything mixed together into a glass evaporator, and then extract a liquid by distillation. During distillation add to the liquor cowslips, 2 handfuls. Johannes Stocker, *Empirica*, book 1, ch. 2, p. 4.

Philip Grüling, whom the learned men value highly, in *Praxis medica*, book 1, p. 53, describes another remedy for paralysis, in which *saffron* is of no little importance.

Or

Take of: Bay-berries, sweet flag,
cypress, calamite storax, of each 1½ ounces

roses, aloes, mastic, myrrh, bdellium, gum ammoniac, alum, of each 6 drachms; cardamom, dill, balaustines, spike, of each 3 drachms; castoreum, *saffron*, of each 2 drachms. Grind what is to be ground, and make a plaster with wax and oil of spikenard to be applied onto the contused parts, from which the weakening started. Feynes, *Medicina practica, De morbis capitis*, ch. 6, p. 159.

For the paralysed tongue:

Take of: Powdered nutmeg, 1 drachm
powdered pellitory of Spain, sassafras-wood, of each 1 scruple
saffron, ½ scruple.

With oxymel make 8 pills. One of these should be kept in the mouth continuously and moved hither and thither. Zecchio, *Consultationes medicinales*, p. 720.

Or

Take of: Turtle doves' blood, 6 drachms
spirits of lavender, black cherry,

primrose, of each 1½ drachms
saffron,
essence of pellitory of Spain made with
spirit of sassafras, of each 1 drachm. Mix.

Let it be marked: Mixtur ein halb Löffel voll darvon auff der Zungen zuhalten ["Mixture, half a spoonful, to hold on the tongue"].

One might wonder whether the miraculous salt, which Glauber, *Miraculum mundi*, § 2 *De usu salis*, discovered, and with which rain water or distilled water, especially spring water, is loaded, and which joins together wine, and vinegar, beer, honey, etc. in a solidity rivalling that of ice, can also induce paralysis?[2]

And in the second place, whether maybe the most esteemed De Bils operated on dogs without the shedding of blood thanks to a similar salt, or rather thanks to his excellent knowledge of blood circulation and the driving of blood from one side to the other? This is left for consideration to everyone's own individual judgement.

III Vertigo

What is vertigo?

Just as air absorbing vapour in the early hours of the morning announces that there will be lots of rain after lunch, so vertigo, especially if it is rather severe and long-lasting, is usually a forerunner of apoplexy, see Galen, *Aphorisms* 2, 17 and 23.

What is vertigo?

Vertigo is the disordered movement of animal spirits, because of which all things are believed to be spinning around in circle. It is caused by blocked circulation, because of the thickness of blood in the arteries of the brain disseminated in the structure of the thick meninges, or by gassy wild spirits that have arrived in the brain with the blood.

The following pills, which strengthen both sensation and head, are specific to eliminate and eradicate completely this condition:

Take of: Choice agaric, 2 drachms
troches of alhandal, 1½ drachms

chebulic myrobalans, 1 drachm; gum Arabic, prepared root of black hellebore, of each ½ drachm; root of asarabacca, true leopard's bane, of each 15 grains; nutmeg, cubebs, cloves, aloes-wood, mastic, of each ½ drachm; car-

2 Glauber's salt is sodium sulphate.

damom, *saffron*, spike, seeds of marjoram, rue, peony, hartwort, ginger, of each 15 grains; crystal salt, saltpetre, of each 5 grains; hepatic aloes, double the weight. With benedicta laxativa, make pills. Dose: 2 scruples to 1 drachm; if used frequently, 1 scruple.

Additionally, Grüling's remarkable plaster of *saffron* is very useful against this condition:

Take of: Native cinnabar,
prepared red coral,
prepared pearls, of each 2 scruples
eastern saffron, 1 scruple
prepared peacocks' dung, 1 ounce

gold leaves, 6 in number. Reduce to an alkool on marble. Dose: ½ scruple or 1 scruple with water of lily of the valley, or rather, chervil-water, which is considered specific for this affection, as are all the other things produced from chervil.

IV Epilepsy

Above all things, earthquakes, clefts, fires, inundations and lightning are dreadful; so Eustratius, 3, *Moralia*. In fact, who does not tremble when the earth shakes? There is no hiding place against earthquakes, no escape, in fact who can escape beyond the earth? Who is not deeply moved by the sight of cleft in the earth? Who is not terrified by fires? When the living become ashes, we cry ash. Who is not silenced upon hearing the terrible word "inundation"? Who is not greatly frightened when the gloomy aspect of the sky foretells impending lightning and thunder, when Pythagoras himself says that the same is done as a threat to those who are in the infernal regions in order to scare them?

What is epilepsy?

He who reflects on the analogy between microcosm and macrocosm clearly sees how in the great world, thunder is produced by the violent clashing of warm and dry exhalations against the sides of a cloud, and from the same exhalations, lightning and thunderbolts are made; and distinctly, how, in the same way, in the little world, epilepsy is produced by air and the boundary of microcosm.[3] Epilepsy is the movement of the whole body, or the non-permanent shaking of some parts. It is caused by blood filling veins

3 Microcosm: here meaning the individual human being.

and arteries in a greater quantity, hindering circulation, and stimulating the excretive capacity; or by spoiled water twitching and disturbing the nerves with its sharpness; or by sharp acid vapours, from the brain or from lower parts, settling in the openings of the celiac artery; or in the pancreatic duct because of an obstruction of mucous tartar impregnated with salino-nitrous particles, and carried to the brain by effervescence when they touch each other; or, finally, intestinal worms biting the bowels and, consequently, causing the whole membranous and nervous system to contract along with it.

Duchesne, *De affectibus capitis*, ch. 35, describes a well-known specific for this affection, to which *saffron* is worthily added:

Take of: Bright mineral cinnabar, ½ ounce.
Rub to an alkool on marble, add:
magistery of pearls,
coral, of each 2 scruples
saffron, 1 scruple
gold leaf (Rolfinck in *Specialis methodus medicinae*, book VII, ch. 5, p. 626, uses ceraunochrysos), 10 grains.

Mix. Make into a powder.

Johannes Naevius (see the letters collected by Scholz, advice no. 309) prescribes the following powder enriched with *saffron* for two-year-old or older children.

Take of: Root of peony, 1 ounce; snake-root, ½ ounce
mistletoe of the oak gathered at the right time, ½ drachm
eastern saffron, 1 scruple
red coral, ½ drachm
root of common dittany, seed of the same, of each 1 scruple; true unicorn's horn, 1 scruple; non-perforated oriental pearls, ½ drachm. Reduce into very fine powder what is to be reduced into powder. Mix and make a powder. Dose: for infants, the amount of one pea or two with the milk of the wet-nurse; for older children, triple the amount with water of betony, lime-flowers, etc.

Grüling, in *Praxis medica*, book I, part I, p. 39, describes another epileptic powder for pharmaceutical use and he does not disdain *saffron*.

Take of: Roots of peony, 1½ ounces; iris, 6 drachms
bistort, 1 drachm
eastern saffron, ½ drachm
whitest sugar, 8 ounces. Mix.
Let it be marked: Pulver vor die schwere Noth ["Powder for epilepsy"].

This is our secret powder:

Take of: Faeculae of peony,
gersa serpentaria, of each 7 grains
magistery of saffron, 2 grains
oil of box by descending distillation, 3 drops.

Secret powder against epilepsy

Mix. For the dose, to be prescribed in tincture of peony, or that water [of peony], or in tincture distilled from wild cherries.

In Stocker's *Empirica*, p. 7, there is also a well known electuary of peony against epilepsy, which contains *saffron* as well.

V Lethargy

Seneca

Those who sleep more than Epimenides are delicate and good at no occupation; especially, as Philip Besold puts it, sluggishness, sleep, every sign of indolence and squalor are banished by the companionship of love.[4] Many slept soundly, and fell asleep in death and death took the place of sleep.

What is lethargy?

This, most of all, is manifest in lethargy, which is an unquenchable need of sleep with fever, forgetfulness of things, and damaged capacity of reasoning, arising because the passage of animal spirits is blocked by very thick, numbing humours or (as others prefer) by very thick nervous liquid.

Excellent mixture against lethargy

Against this affection, as well as against other affections inducing sleep, *saffron* is especially recommended by Freitag, *Aurora medica*, p. 502 et seq. The following mixture of three ingredients is excellent. Premising what is to be premised, it vanquishes excrementitious humours stagnating in the head, and increases animal spirits.

Take of: *Essence of saffron*,
of castoreum,
tincture of silver, of each 1 drachm.

Mix.

Dose: 10 or 12 drops in water of sassafras-wood, or other convenient water.

4 Epimenides (5th or 6th century BC Greek seer and philosopher) famously fell asleep for 57 years whilst tending his father's sheep.

VI Amnesia

If knowledge is the perpetual companion of memory, it is also an integral part of intelligence, since the experience of many things is required, as the Philosopher writes in *Metaphysics*, 2. And as an ancient poet wrote on wisdom:[5]

> *Usus me genuit, mater peperit memoria*
>
> ["Experience begot me, memory was my mother"]

Certainly, without memory no-one can be called wise, no-one experienced.

So, Emperor Claudius was deemed a log, as, after he had his wife Messalina killed by a spy following a weak deliberation (because she was involved in all sorts of adulteries having trodden underfoot the pact of the nuptial bed), reclining at dinner one hour later, but forgetful of the deed, he asked what happened to his wife that she did not come to dinner? Suetonius, *De vita Caesar.*

What is amnesia?

Amnesia is an impairing or a total annulling of the recollection of things done or heard. It is caused by a deformed conformation of the brain itself or by its excessive moisture, by beating, by old age or by long illnesses of the brain, by philtres, sleeping drugs incautiously used internally or applied externally, or finally by a very heavy diet relinquishing thick and sooty spirits.

Sennert recommends the following pills for memory, which also strengthen the head, and he does not want them without *saffron*:

Take of: Aloes, agaric, of each 2 drachms
crystal salt, ½ drachm
cloves, cinnamon, nutmeg, mace, aloes-wood, frankincense, mastic, choice myrrh, cubebs, *saffron*, wormwood, bay-berries, juniper-berries, sweet flag, of each ½ drachm. Make into powder and mix with sweet wine and make the pill mass.

Lauremberg's aqua magnanimitatis, very much praised by him, is also of great help for amnesia. And the author did not deem it unbecoming to add *saffron* to his aqua, like the easy and remarkable remedy described by Zacuth of Lisbon, *Praxis historiarum*, pp. 219–220, in which the virtue of *saffron* matches its pleasantness.

---

5 The "ancient poet" is Afranius.

The ointment of Charles, Duke of Burgundy, which is purchased at the price of one thousand florins and is believed to procure perpetual memory of things, is recommended with exceptional praises. This is its recipe:
Take of: Leaf-juices of balm, basil,
flowers of tamarisk, lilies,
primrose,
lavender, rosemary,
borage, broom, of each 2 ounces
aqua vitae, water-lilies, roses, violets, of each 1 ounce; cubebs, cardamom, grains of paradise, yellow sanders, carpobalsam, iris, *eastern saffron*, savoury, peony, thyme, of each ½ ounce; liquid and calamite storax, opopanax, bdellium, galbanum, gum of ivy, ladanum, of each 6 drachms; roots of long birthwort, peony, oils of turpentine, nard, costus, juniper, bay-tree, mastic, of behen, of spike, of each 5 drachms. Grind what is to be ground, and mix everything. Distil with an alembic according to the degree of the fire, and let the water and oil be separated. For the first two months the use will be this: anoint nostrils and channels of the ear with an amount of ointment equal to the size of a pea; for the following two months use every third day; for another two months use twice a week; then once a week; afterwards once in a fortnight until the year is completed; then, in the following years use once every two months.

VII Incubus

The word Incubus pertains to spirits. The Epicureans laughed that such thing existed and regarded as fiction and mockery the things by which children, and ignorant and ingenuous men were frightened. Among them are included the Sadducees too, who denied the existence of angels and spirits, see *Acts of the Apostles*, ch. 23. However, there are several faint-hearted people who, when they hear the earth pulsating in the horse stable during the dead of night, immediately tremble and sweat, thinking: I don't know what ghost is wandering about in the house during the still night. Likewise, when a woodworm gnaws a wooden wall, others say that they heard a ghost lightly hitting an anvil with a mallet. Do spirits exist?

In contrast though, men of as strong and noble character as the valorous Scythian people (from whom the Turks descend) are never haunted by ghosts.

Nonetheless, the things which follow prove their existence, as now and then they bring violence, sometimes death.

Lares, Lemures, and Manes are counted among spirits.

In fact, the Gentiles (see Augustine and other Church Fathers) believed that the souls of men became spirits, and they called these Lares if they were good, Lemures if they were evil. And when they were uncertain whether they deserved well or ill, they called them Manes.

Nonius names Lemures nocturnal ghosts and terrifying apparitions and beasts; and in the same way as the Lares are considered good, the Lemures are considered evil.

Others conceive Lemures as shadows, and shadows as the souls of the dead.

The word Manes derives from *mane* ["morning"], which is good; whence the gods Manes, just as good and favourable. Soothsayers said that the Manes were celestial deities as well as infernal deities, because they believed that all things originated on account of them.

Others claimed that the Manes were assigned to bodies immediately after conception, and that, in order not to abandon the bodies, they live in burial grounds even when the bodies have consumed. Therefore, those who destroy burial sites are believed to dishonour the Manes.

Others call the Manes tutelary deities, and, of them, every single person is allotted a good one and an evil one. The good one is needed to strengthen the spirit of men towards good, the evil one to preside over the man's flesh and sexual desire; and the good overseer will fight against the enemy and the flesh. Man, however, caught in the middle between the two contenders, and at the mercy of his own judgement, helps the one he wishes to win.

Here also belong witches, namely apparitions of demons, who, assuming the shape of beautiful women, devour children and young people whom they have enticed.

And the spectre Empusa, which throws itself on miserable people, changes many shapes and usually appears at noon. And also Dicelon, so called because it frightens, in Greek Hecatea from Hecate or Proserpina, who is believed to let loose similar terrors at night.

Not surprisingly these spectres appears more often during the night than during the day. In fact, in the Holy Scriptures they are called princes of darkness, because they escape the light of God.

I say nothing of the Telchines, men using sorcery, undoubtedly possessed with a spirit, and insane. Likewise, I omit Panes, Faunes, Satyrs, Sileni, Onocentaurs, Onosceli, Hippotaurs, Sphinxes, Scyllae, Harpies, Tritons, Nereids, and Mermaids.

Now I will talk about those reviling demons named Ephialtae, Hyphialtae, Incubi and Succubi.

And the conjecture that Melusine was among their number is not incongruous.[6]

Thus, Mirandola writes about a 75-year-old who, for more than forty years, slept with a spirit in the form of a woman.

Similarly, Cardano claims that he knew someone who, in like manner, slept with a spirit for more than forty years, and called it Flora and talked about this at length.

In Smirtz, a village under the domain of Wratislaus von Bernstein, a woman who was visited by an Incubus gave birth to a horrible monster, and, by the order of the Senate of Olomouc, this occurrence was broadly divulged in Moravia.[7] Bodin, in his *Daemonomia*, mentions a woman, Johanna Harvilena, who, having known an Incubus several times, was asked by it whether she wanted to be made pregnant. However, as I do not want to hold the kind reader longer, I will pass over the question whether the semen of a demon, ejaculated without a human body, and received repeatedly, can undoubtedly make a woman pregnant, and I will go back to the Incubus. The Jews write that these, as well as the other apparitions already enumerated above, were created from Adam's seed in those hundred and thirty years during which he abstained from his wife, after Cain had killed his brother.

According to a singular remedy, seven circles are to be made around a grave, so that they do not go near a corpse to possess it, and do not use it to do harm.

In the books of the ancient physicians Actuarius and Alexander, many remedies are found, which were used against similar spectres.

What is an Incubus?

Now, these things are said to happen by the work of a demon; the same things, however, often also befall entirely by nature as the Medical School confirms. They explain that Incubus is the disappearance of movement, voice and respiration originating because the afflux of animal spirits into the nerves is blocked; or because the stomach is too full of food and drink and, thus, hinders the movement of the diaphragm; or because the chyle does not run quickly enough through the septum of the heart and the pulmonary blood vessels into the left ventricle of the heart and, on account of this, into the whole body.

6 Melusine: a female spirit connected with fresh water.

7 Smirtz may be Smržice, in German Smrzitz in Mähre, a village in the Olomouc region of Moravia (eastern part of the Czech Republic). Olomouc itself was the historical capital of Moravia. Wratislaus von Bernstein, also called Vrtatislav von Pernstein, lived 1530–1582.

In a manuscript codex of Hippocrates' works one reads that an antidote for those frightened by Lemures and possessed by demons, and lunatics with foam, will benefit sufficiently if it also contains *saffron*:

Take of: Cypress nuts, 50 in number
fat dried figs without seeds,
raisins cleaned of pips, of each 12 ounces

ladanum, best storax, myrrh, asphaltum, bdellium, root of asphodel, root of elecampane, of each 7 ounces; cardamom, 5 ounces; savin, 4 ounces; leaves of costus, spike, roses, juniper-berries, of each 3 ounces; schoenanth, cassia, mastic, of each 2 ounces; *saffron*, English galingale, lovage, of each 1 ounce; best wine and honey, as much as suffices.

Crush the cypress nuts and let them macerate for three days in the best wine. Boil over a slow fire for a while until the wine is consumed, and strain. Then in 2 pounds of wine, pound the ground pulp of the figs at length with the wine, and add 12 pounds of honey. Boil again until the honey becomes thick, and strain through sack-cloth. Then add raisins and pulp of dates, and pound thoroughly in a mortar. Then mix in the mortar storax, ladanum, turpentine pounding at length with a little honey, and after this, carefully put the powders together and store up in a glass. Dose: 2 or 3 drachms.

By experience we learned that the following pills protect from Incubus, when it has natural causes:

Take of: Best aloes, 2 scruples
smaragdine treacle, ½ scruple
extract of saffron, 2 grains.

With tincture of peony flowers, make 21 pills as a dose.

VIII Catarrh

Very often states and rulers improvidently foster snakes in their bosom and rats in their bag, and offer the weapon by which they will be destroyed.

What is catarrh?

Among the most important parts of the human body is the brain. The latter very often produces excrementitious humours until, the bolts having finally been loosened, just as from Pandora's box or from another Trojan horse, they come forth to produce a myriad of diseases, and present to the irritated archeus the weapon with which it can prepare many battles and take over the control of our body. But who is the man who does not know catarrh? Catarrh is a discharge of thick or sharp or thin humours through the orifices of the cribriform plate of the ethmoid bone and the four membranous side conduits around the sella turcica into the adjoining parts; or a stagnation of viscous particles of the same very cold blood circulating through the lungs, which frequently obstructs the thin arteries that must be a thoroughfare, thus causing a suffocation.

Therefore, the following hydromel is extremely beneficial against the formation of pituitous and viscous humours flowing from the brain into the lungs or stagnating in the lungs.

Take of: Herb of wild thyme collected at the right time,
coltsfoot, sage,
betony, mountain cudweed,
scabious, of each 1 handful

roots of elecampane, coltsfoot, orris, of each 1 ounce; round birthwort, 6 drachms; flowers of scabious, betony, elecampane, of each 2 pugils; best clarified honey, 1 pound; water, as much as suffices. Boil. At the end add: *eastern saffron*, 1½ drachms; cardamom, cubebs, of each 1 drachm. Add to the colature essence of sassafras-wood, 3 drachms. This is an alexipharmic against all catarrhs. Mix.

The following mixture of *saffron* is excellent for cutting into the sticky matter:

Take of: *Our oxymel of saffron*,
syrup of horehound, of each 6 drachms; water of hyssop, 2 ounces; fennel, betony, of each 1 ounce; *pectoral elixir of saffron*, 1 drachm. Mix.

Or

Take of: The *compound sugared oil of saffron*, 1½ drachms
species diaireos, 1½ drachms
faeculae of cuckoo-pint, ½ drachm.

Mix. Make an eclegme.

Let it be marked: **Brust-öffnungs Saffran** ["Chest-decongestant of saffron"].

The recipe of the *compound sugared oil of saffron* is this: *Compound sugared oil of saffron*

Take of: Powdered whitest sugar three times besprinkled with *vinegar of saffron*, and as many times dried, ½ pound. Pour over this spirit of wine, 1 pound, in which stoned dates, pitted raisins, of each ½ ounce; roots of orris, elecampane, of each 3 drachms; *Austrian saffron*, 1 drachm; have been macerating for three days. After carefully straining the heated spirit of wine from the aforementioned dry ingredients and letting it be consumed by fire, add to the thick oil which is left: the *asthmatic spirit of saffron*, 1 drachm. Mix.

Take of: Spermaceti, 2 drachms
powder of saffron, ½ drachm.
Mix. Make powder.

Let it be marked: **Pulver gegen den Stockfluss auff 4. mahl in Hysop-wasser oder warm Bier zugeben** ["Powder against catarrh to be given 4 times in hyssop-water or warm beer"].

For strengthening the head and drying the catarrh in there, see the skull-cap described by the philosopher Grüling in *Praxis medica*, book I, part 2, p. 88, which contains *saffron*.

The famous Sennert, Phoenix of the German Physicians, praises these orbicules against thin and salty catarrh:
Take of: Poppy-heads, 7 in number
carob, 1 ounce
root of liquorice, 5 drachms
root of iris, 1½ drachms; raisins, 3 drachms. Boil according to Art, in a sufficient quantity of water, down to the third part. In the strained decoction, dissolve:
sugar, 4 ounces
extract of liquorice, 1 scruple
saffron, ½ scruple.
Make orbicules according to Art.

The famous Franz de le Boë Sylvius considers these tablets serviceable against thin and sharp catarrh:
Take of: Seeds of henbane,
white poppy, of each ½ drachm
white frankincense, red myrrh, of each 1 scruple
best saffron, ½ scruple
inspissated juice of liquorice, 1 drachm; marshmallow, 3 drachms; gum tragacanth dissolved in rose-water, 1 drachm. Make tablets to be frequently held in the mouth.

The following pills are for curing catarrh of the same kind:
Take of: Calamite storax,
juice of liquorice, frankincense,
red myrrh, Thebaic opium, of each ½ drachm
saffron, 1 scruple.
With syrup of wild poppy, make a pill-mass.

Moreover, the excellent Johannes Crato, Chief Physician to three Emperors, prescribes this liniment for the anointment of the chest:

Take of: Oil of sweet almonds, 1½ ounces
goose fat, 1 ounce
saffron, 1 scruple. Mix.

Scholz in *Consiliorum medicinalium*, p. 62.

IX Agrypnia, or Wakefulness

The Egyptians justly drew the hieroglyphic of wakefulness and watchfulness as the head of a lion, because the lion alone, among the animals with crooked claws, sees as soon as it is born. Pierio, vol. VI, book XXXII, p. 118.

This is confirmed by Alciato in his *Emblemata* XV:

Est Leo, sed custos, oculis quia dormit apertis,
Templorum idcircò, ponitur auto fores.

["The lion is indeed a watchman, because it sleeps with open eyes,
For this reason it is placed in front of the doors of temples."]

And Homer, honour of the Greeks, said that, in the same way as nature bestowed this gift upon the king of animals, thus this very gift particularly suits rulers. However, physically speaking, if sleeplessness exceeds measure, it is harmful. Hippocrates, *Aphorisms* 3, p. 2. Namely, three sleepless nights weaken the body exceedingly, as Ovid writes in his *Ars amatoria* I:

Attenuant juvenum vigilatae corpora noctes
Curaque & immenso qui fit amore dolor.

["Sleepless nights and concern and the pain of a great love
weaken the bodies of young people."]

Agrypnia is the relaxation of the common sense and the external senses inconsistently with the established rhythm and time, and [is characterised by] motor, and lasting, energy. It is caused by confused or restless animal spirits, by pain, or by the rejection of sweet evaporations, or of moist temperament. What is agrypnia?

Its cure is *water of saffron* distilled from the flowers with or without the threads, taken two hours after dinner, up to 4 ounces. Mattioli, book I, ch. 16, p. 78.

Or instead:
Take of: Excorticated sweet almonds, 2 drachms
the 4 greater cooling seeds, excorticated, of each 1 scruple.
With *distilled water of saffron-flowers*, 3 ounces, make an emulsion according to Art. Sweeten with syrup of poppies for a dose.

If the agrypnia is very strong:
Take of: *Conserve of saffron-flowers*, 2 drachms
the strengthening anodyne powder, 10 grains.
Mix.
Let it be marked: **Abend-zucker auff einmahl** ["Evening sugar, once"].

And the following pills are no less effective:
Take of: Bole armeniac, spikenard,
myrrh, frankincense, of each ½ drachm
opium, *saffron*, of each ½ scruple.
With syrup of poppy, make pills the same as lupins [seeds]. Dose: 1 drachm.

Neither without praise is Daniel Angelus' sleeping powder, described in the *Appendix de compositione medicamentorum*, p. 22:
Take of: Seeds of henbane,
white poppy,
rind of mandrake, cassia, of each 2 drachms; schoenanth, 1 drachm; Thebaic opium, spagyrically-prepared, 1 scruple; mace, *best saffron*, of each 1 scruple; white sugar candy, ½ ounce. Mix. Make a powder. Dose: 1 or 2 scruples.

Additionally, a wreath of *saffron-flowers* applied to the head prevents wakefulness.

Or apply to the head this excellent epithem when sleeplessness is caused by a headache.
Take of: Sweet almonds,
walnuts, of each ½ ounce
eastern saffron, 1 scruple
white poppy, 1 drachm
water of roses, *saffron*, of each 1½ ounces; wine vinegar, 1 ounce. Mix. Make an epithem.

Or
Take of: Oil of nutmeg by expression, 2 ounces
yellow amber, 1 scruple

extract of saffron, 6 grains
camphor, 3 grains.
Mix. Make a balsam. Werner Rolfinck in *Consultatio*.

Or
Take of: Oil of poppy by expression, 1 drachm
oil of peach-kernels by expression, ½ drachm
extract of opium, 3 grains; *extract of saffron*, 5 grains
oil of nutmeg, as much as suffices. Make a liniment.

Or
Take of: Oils of violets, water-lilies,
ointment of poplar-buds, of each 2 drachms
opium, *saffron*, of each 3 grains. Mix.
And with this anoint temples and nostrils.

In Padua, Wolfgang Gabelkover saw a very famous and expert doctor, one Albertino Bottoni, use this very remedy, thanks to which he too very often obtained honour. *Curationum et observationum medicinalium, curatio* 5, p. 10.

Stocker in *Empirica*, book 1, ch. 7, p. 17, recommends this ointment by Bartolomeo Montagnana against particularly resilient sleeplessness:
Take of: Oil of poplar, 1 ounce; violets, ½ ounce
henbane,
rind of mandrake, of each ½ ounce
saffron, ½ drachm.
Mix. Make an ointment. With it, anoint the temples, nostrils, wrists, and soles of the feet.

When sleeplessness troubles with malignant fever:
Take of: Alabastrine ointment,
ointment of poplar-buds, of each 6 drachms
saffron dissolved in water of water-germander, 1 scruple.
Mix. The famous Thomas Bartholin, *Cista medica*, p. 491.

X Phrenitis

The masses are not governed easily, and their impudence and fierceness are so great that against no-one they rise more than against those to whose command they must obey.

Not differently, the delirious person is taken by a raging frenzy without restrain, is disagreeable, cruel, and unrestrained, and not uncommonly rages without control against physicians and bystanders, to whose authority and indulgence they are committed, being full of indignation. Physicians should approach patients of this sort cautiously and circumspectly: in fact, past events give true warnings for future events.

What is phrenitis?

Phrenitis is a continuous delirium with high fever. It is caused by an inflammation of the membranes of the brain, by warm blood boiling up and pouring from the greater arteries into the smaller, and from the smaller into the smallest into the porosity of the thin membrane surrounding the brain, and then, after being denied a way out through the hollows in the veins, producing vibratory motions and nodes.

After performing a venesection, which is absolutely necessary, among the things that immediately cure phrenitis in its initial stage is the following draught:

Take of: Laudanum, 3 or 4 grains
water of bugloss, 1 ounce
tincture of saffron,
borage, of each 6 drops
essence of coral, 2 drops
the 5 precious stones, of each 1 grain.
Mix. Make a draught.

Or, when there is a fear that narcotics might not induce sleep:
Take of: *Water of saffron-flowers*, 2 ounces
pimpernell,
water-lilies, of each 1 ounce
the 4 greater cooling seeds, of each 2 drachms; white poppy, 3 drachms. Make an emulsion, add:
syrup of oriental pearls, 3 drachms;
true oil of pearls, 6 drops.
Mix for one dose.

Or
Take of: *Water of saffron-flowers*, 2 ounces
sugar of lead, 4 grains.
Mix.

When, on account of sleeplessness, the patient shouts out, Zacuth of Lisbon, *Praxis historiarum*, p. 680, uses these very pills of opium and *saffron*:

Take of: Opium, 6 grains; *saffron*, 5 grains
castoreum, 2 grains.
With juice of henbane, make pills, which, put in soup, are to be taken often.

XI Mania

Just as the fiercest bulls are tamed with a little rope, even very thin, tied to their right knee (this has been tested in Rome by one Graculus under the pontificate of Clement VII) thus, their common sense being bound, mad people are very often tamed and made quiet and calm with a light medicine, which restrains and softens the ardour and the violent rage.[8]

Mania is, in fact, a delirium without fever, with rage and daring, caused by a fiery and almost burning disposition of the animal spirits because of warm, atrabilarious humours or because of a philtre. What is mania?

While every hope of healing depends for the most part upon bloodletting and limiting the wakefulness, internally, *water of saffron-flowers* administered everyday up to 2 ounces with 6 or 7 drops of tincture of silver is beneficial.

Externally, the following ointment will be good:
Take of: Alabastrine ointment, ½ ounce
ointment of poplar-buds, 2 drachms
opium, 8 grains
saffron, ½ scruple
distilled oils of dill, chamomile, of each 2 drops. Mix. Make an ointment.
Balthasar Timaeus von Güldenklee, book I, *De affectibus capitis*, p. 18.

8 Aelian (book IV, no. 48) writes that nothing can stop an angry bull except that "a man may bring him to a halt and stay his onrush if he tie a scarf round his own right knee and face the Bull." Clement VII was Pope between 1523 and 1534.

CHAPTER IX

Diseases of the Eyes

1 Ophthalmia

Does bewitching exist?

It is common opinion that bewitching is performed in three ways: with a look, with the voice, and with contact. Most people, however deny that bewitching happens if not because of a spell or sorcery.

Balthasar Timaeus von Güldenklee, *Casus medicinales, De morbis ex incantatione*, p. 327 et seq., and other practitioners, enumerate sorcery of this kind.

But what is there to think about ophthalmia, about which a poet writes:

Dum spectant oculis laesos, laeduntur & ipsi,
Multaque corporibus transitione nocent.

["While the eyes gaze upon the injured, they become injured themselves,
and many things harm the body through contagion."]

Does ophthalmia work in the same way as witchcraft? It is beyond doubt that healthy effluvia emanate from healthy bodies and sickly effluvia from sickly bodies. In fact, who does not know that little old women contaminate delicate infants? That menstruating women stain mirrors? That men with a contagious fever transmit it to another body through clinging, wormy effluvia? The same happens with ophthalmia; when the effluvia exhalating from a sick eye reach a healthy eye, since they are different, by hurting one another because of their opposite nature, they cause an inflammation in the eye.

What is ophthalmia?

Ophthalmia is an inflammation of the tunica adnata with pain, a burning sensation, and redness, caused by boiling hot very serous blood swelling out the veins and arteries of the adnata, or by a sting, a fall, a contusion, or by smoke, etc.

Against this condition, in addition to venesection, which is extremely useful and is exceptionally praised by Galen in *De curanda ratione* as it alone cured the most violent inflammations within the hour, we recommend these remedies with *saffron*.

 | DOI:10.1163/9789004435292_013

Take of: Seed-mucilages of psyllium, quince,
extracted with rose-water and poppy-water, 1 ounce
egg whites, shaken, ½ ounce
camphor, *saffron*, of each 5 grains.
Mix. Make a collyrium to be poured into the eyes by drops. Sennert.

Or
Take of: Waters of plantain, tormentil,
roses, cornflowers, of each ½ ounce
egg white, 1 in number
saffron, alum, of each 5 grains
Let it be marked: Augenwässerle ["Eye water"]. This is Timaeus' panacea.

When there is pain:
Take of: Core of a loaf of the best bread, ground, 1 ounce
sweet apple, ground, 3 ounces
woman's milk, 1 ounce
sugar, 2 drachms; camphor, 15 grains
saffron, 5 grains.
Apply with two linen cloths. For contusions and wounds of the eyes as well.

Or
Take of: Pulp of roasted apple, ½ ounce
fresh white breadcrumbs, 3 ounces
powdered saffron, 1½ scruples
fresh woman's milk,
rose-water, of each equal parts.
Mix. Make a cataplasm. Diemerbroeck.

Take of: Breadcrumbs of a loaf made of the finest wheat flour, 3 ounces
flowers of chamomile,
melilot, of each 1 pugil
rotten apple, 2 ounces
egg yolks, 2 in number
seed of fenugreek, 1 ounce
saffron, ½ drachm.
Boil in cows' milk and make a cataplasm. Morel, *De materia medica*, p. 504.

Or
Take of: Water of liquorice,

fresh woman's milk, of each ½ scruple
camphor, *saffron*, of each ½ scruple.
Mix. Make a collyrium. Möbius, *Anatomia camphorae*, ch. 15, p. 71.

The citrine collyrium against severe pain is thus made:
Take of: Burnt tutty quenched and washed in woman's milk, ½ ounce
washed ceruse, 1 ounce
gum tragacanth, 1 drachm
saffron, 2 drachms; opium, ½ drachm
rainwater, as much as suffices.
Use this with egg white. Felix Platter, *Praxis medica*, vol. 2, p. 274.

Additionally, Hippocrates' remedy (*Epidemics*, book II, section 5) can be used to stop discharges.

He says to pour water upon the head of the patient and to apply this very remedy to the eyes.
Take of: Thick frankincense, ceruse,
myrrh, of each 3 drachms
tutty, *saffron*, of each 2 scruples & 4 grains.
Wash the ceruse and the tutty, and make a collyrium. Add a little wine in the mortar and let it dry in the sun, then dissolve in milk or whey, and apply to the eyes by drops.

Against the condition of ophthalmia, as soon as the humour is not very sharp and the tears shed from the eyes seem fewer, apply remedies that dissipate it at once, for instance:
Take of: Sarcocolla nourished in asses' milk, 1 drachm
Socotrine aloes, 2 scruples
prepared tutty, ½ drachm; *saffron*, 5 grains
or a little more
seed-mucilages of psyllium, quince
extracted with rose-water, [of each] 1 ounce.
Mix. Make a collyrium. Some drops of woman's milk can be mixed to it when it needs to be used. Feynes, *Praxis medica*.

When the condition is as above:
Take of: Nourished sarcocolla, 1 drachm
aloes, 1 scruple
saffron, ½ scruple
spike or musk, 5 grains

water of eyebright, or sanicle,
or elder, of each 2 ounces.

Mix. Make a collyrium. If you wish, add mucilage of fenugreek, 1 ounce. But before, foment the eyes with a decoction of fenugreek from rose-water or white wine. Feynes, *Practica medicina*, p. 209.

When it is about to start disappearing and when it is necessary to get rid of it, Rhazes' citrine powder with sarcocolla will be helpful:

Take of: Sarcocolla, 5 drachms
aloes, *saffron*, of each 1 drachm
myrrh, ½ drachm; lycium, 1 drachm.

Grind to very fine powder with a grindstone.

When it is disappearing, and the discharge and pain start to fade, to dispose of the last of it:

Take of: Sarcocolla nourished in milk, 3 drachms
aloes washed in sanicle-water, 1½ drachms
saffron, 1 scruple. Make a powder.

From this, with eyebright-water, make a dry collyrium, which is to be used in fennel-water. Sennert.

Take of: Frankincense, aloes, of each ½ drachm
sarcocolla nourished in woman's milk, 1½ drachms
saffron, ½ scruple
seed-mucilage of fenugreek, ½ ounce
waters of fennel, eyebright, of each 1½ ounces.
Mix. Make a collyrium. Rivière.

Saffron, being a solar plant, is indeed extremely appropriate for the eyes, especially if, by means of pharmaceutical Art and the influence of the sun, an eye-water is prepared with it, see Hiebner, *Mysterium sigillorum, herbarum et lapidum*.

11 Cataract, or Opacity of the Cornea

The light is sweet and it is delightful for the eyes to behold the sun. In fact, much of reality is hidden from the blind, as the sense of sight shows us the diversity of many things. However, in the same way as courage grows from despair to those who are besieged, and when there is no hope, fear takes up

Ecclesiastes 22; Seneca, *Tragedies*; Aristotle, *Metaphysics*, book I.

weapons, thus, with a cataract, when all the other remedies have been tried, one turns to an iron or a needle for a cure.[1]

What is a cataract?

A cataract is the clouding of the crystalline lens caused by thick viscous humours accumulated in the eye, and transformed into a thin film with the heat.

The following water is effective:

Take of: Celandine-water, distilled from the juice, 2 ounces
honey-water, distilled from the comb, 1 ounce
crocus metallorum, 1 drachm
saffron, ½ scruple.

Let them be infused for three days, then make a colature with a filter, and to this add:

white sugar candy, 2 scruples.

Mix. Make a collyrium.

Also, the distilled water described in Feynes, *Medicina practica*, book II, p. 218, is beneficial for opacity of the cornea, albugo, nubecula, a scar, pterygium, as well as the distilled water described by Gulielmus Locellus, and after him, by Rivière, ch. 4, *De suffusione*. This water, after the universal remedies, prides itself with having cured many who did not enjoy any use of light, and it contains *saffron*.

Likewise, Hippocrates' famous remedy against cataracts, for which see Rodrigues de Castro's *Consultationes medicae*, p. 145.

For this purpose the ointment called crocomagon is also effective. Bartholomaeus Anglicus, *De proprietatibus rerum*.[2]

Alternatively, first foment the eye with dry red wine, which stops the discharge with its astringent property, while dissipating and softening because of its alcoholic content. Then, when the disease has progressed further, apply the following cataplasm:

Take of: Meal of fenugreek, 1 ounce
aloes, ½ ounce
saffron, 1 drachm.

---

1 Cataracts were famously treated by couching, where a needle was inserted into the eye to dislodge the lens into the vitreous humour, often by itinerant quack doctors. Oculist to the Elector of Saxony, Georg Bartisch despairingly wrote (in his *Ophthalmodouleia* of 1583: translated section quoted by Chan, 2010) that there was no "lack of old women, vagrant hags, therica sellers, tooth-pullers, ruined shopkeepers, rat and mouse catchers, knaves, tinkers, hog-butchers, hangmen, bum-bailiffs, and other wanton good for nothing vagabonds ... all of whom boldly try to perform this noble cure."

2 The ointment crocomagma is possibly the same as crocomagma on p. 82.

Make a very fine powder, pour in white wine, in which crocus metallorum has been infused, and make a cataplasm to be applied at bed time. Rivière.

In cases of hyposphagmata, when the humours of the eyes are bruised by a blow or a fall, but without any eruption, Feynes, *Medicina practica*, book II, ch. 22, p. 225, commends the following remedy:

Take of: Frankincense, *saffron*,
myrrh, of each [no quantity given].

Pulverize these accurately, and put them in the eye as a powder, and a bloodshot eye does not need to be cured with other remedies.

III Phlyctens

If the opinion of the Manicheans were true, that a good god is the creator of good things and an evil god of evil things, then, given the different origin and purpose of either, you would rightly say that the good god is the creator of pearls, the evil one of blisters.

As, in fact, when the time is right for conceiving, shells are as thirsty for dew as for a husband. Then, gaping with desire for this, when the moon drops are extremely fluid, they swallow this much desired liquid with a gaping and are impregnated and produce large pearls, which are called among other things 'blistery excrescence of the shell'.

What are phlyctens?

Thus, on the other hand, very often pernicious pustules like pearls originate from a serous, sharp and pungent humour insinuating itself in the tunica adnata or between the adnata and the cornea.

Besides the universal remedies, this remedy with *saffron* is unparalleled to remove phlyctens:

Take of: Seed-mucilages of psyllium, quince,
fenugreek, extracted in rose-water, 2 drachms
rose-water, 1½ ounces
flowers of elder, ½ ounce
saffron, ½ scruple.

Mix. Make a liquor. Apply to the whole eye with soft linen cloths.

IV Ungula

He who desires peace should prepare for war: in fact, the treacherous neighbour is an enemy. The eye experiences this from a close ungula, and if it does not prepare a war against it early enough with a medicine, it will be very soon surrounded by Cimmerian darkness.

In fact, the ungula of the eyes is a little nervous membrane beginning from the canthus of the eye, very often from the side of the nostrils, sometimes also from the side of the temples, which at times extends to the pupil, and obstructs the sight.

The following, among many, consumes the ungula in the space of four days, and lifts the pannus:

Take of: Woman's milk,
saffron, gall of goat, of each as much as you please.

Let it boil up a bit in an eggshell; instil 2 drops in the eye. Van Foreest, *Observationes*, book II; Grüling, *Medicinae practicae*, book I, part 2, ch. 8, p. 112.

V Aegilops

A military skill is cleverly taught by those endowed by it, which uses tunnels to overcome defensive walls, with this use in particular, that, when the tunnels have reached the foundations, they hastily stop the destruction of the walls, thus Flavius Vegetius, ch. 24 *De bello*.

What is aegilops?

The eye too is attacked by aegilops with tunnels or fistulas, by which an easy destruction of the eye can follow, should there be no vigilance or industry against them.

In fact, a fistula is an affection of the larger canthus of the eye provoked by the excessive influx of serous and sharp humours causing the larger gland of the eye to suppurate.

Against this affection, premised the turning away of the affluent humours, Samuel Hafenreffer, *De cutis affectibus*, book 2, ch. 14, p. 306, prescribes this:

Take of: Gum ammoniac, sarcocolla, of each 5 scruples
frankincense, 3 drachms
saffron, 2 scruples.

Reduce all these things, with the addition of fenugreek mucilage, to a collyrium.

VI Procidentia of the Eye

There is a common saying: one's own house is the best and the most comfortable house. On this account, Aglaus, a poor old man, was declared by the oracle of Delphi happier than King Gyges of Lydia, because he had never left the boundaries of his farm, where he was born.[3]

It is rather inconvenient for the eye too, when it departs and is driven away from its orbit.

Procidentia is the prominence of the eye or its bulging out of the orbit because of humours enlarging it in mass and breadth.

What is prociden-tia of the eye?

Now, it is necessary to deal with the underlying causes of this.

First, foment well the eyes with steam or a lukewarm liquor taken with a sponge from a decoction of chamomile, melilot, or fenugreek. Then apply the following collyrium to the bulging eyes like a cataplasm, with very soft silken wool.

Take of: Virgin honey, 1 drachm
saffron, ½ drachm
myrrh, frankincense, of each 1 scruple.

Reduce to very fine powder; mix them with honey, and make an ointment. François Joël, *Practicae*, book II, p. 219.

VII Swelling of the Eyelids

He who sits in the dark and does not see the light of the sky is imprisoned in an ugly prison. In fact, light comes from good and is the image of goodness.

Dionysus the Aeropagite, *De divinis nominibus*

Not differently, the eye surrounded by swollen eyelids, as if encircled by a wall of darkness, wallows unhappily in the dark cave of the body like another Homeric cyclops after he was blinded.

The swelling of the eyelids is the lifting of the eyelids caused by pituitous humours, exhalations, spiders' or wasps' bites, wakefulness, excessive rubbing of plants such as spurge, laurel, etc. And the eyelids of infants swell the most.

What is the swelling of the eyelids?

To remove such swelling, François Joël, *Practicae*, book IX, section 2, p. 505:

Take of: Root of celandine, red rose, of each 2 drachms
myrrh, *saffron*, of each 1 drachm.

Reduce to very fine powder and add woman's milk, as much as suffices.

For an epithem, which is to be applied on the swollen eyes with linen cloths.

3 King Gyges reigned 716 BC–678 BC in the kingdom of Lydia (today's western Asia Minor).

CHAPTER X

Diseases of the Ears

I Deafness

Gulielmus Ader

It is well known that deaf people abound in maritime places. Hence, deaf people were cured by Christ our Lord especially in the area of Tyre and in the maritime regions of Sidon and around the Sea of Galilee.

Cicero in the *Somnium Scipionis* is convinced of the astonishing, rapid motion of the globes of the universe, explaining that ears filled up with noise become deaf, as in us there is no weaker sense [than that of hearing]. This seems to be confirmed by people dwelling by that place where the Nile falls from very high mountains into the Catadupa.[1] In fact, they are deprived of the sense of hearing because of the great extent of noise.

Neither, perhaps, is the one who attributes the cause of this to the humidity of those places, wrong. In fact, as boreal dispositions strengthen the sense of hearing, thus austral dispositions weaken it. Hippocrates, *Aphorisms*, book III: 17.

What is deafness?

Deafness is the perception of no sound. It can be a birth condition or be caused by the breaking of the tympanic membrane or by the obstruction of the auditory nerves because of a discharge.

When the cause of deafness is an obstruction:

Take of: *Saffron*, cloves, of each ½ drachm

Put in juice of sow-bread, and the mixture in the ears 1 drop per time, warm. It is a panacea for this, see Jean Prévost, *Selectoria remedia*, p. 84.

Oswald Grembs in *Arbor ruinosa*, p. 194, recommends another remedy with *saffron*.

Or

Take of: Aloes, 1 drachm
mastic, seeds of rue, mustard,
saffron, cloves, of each ½ drachm
clarified honey, as much as suffices.

It makes a liniment which is effective against a long-lasting affection of the ear and restores the sense of hearing.

1 Catadupa: cataracts of the Nile, between Aswan and Khartoum.

 | DOI:10.1163/9789004435292_014

Or
Take of: Minderer's or Mynsicht's acovistic spirit,
of *saffron* (this has great loosening power), of each ½ ounce
essence of gall of man or partridge, 2 drachms.
Mix. Insert, tepidly, into the obstructed ear with cotton.

Or
Take of: oil of grubs found in the bark and wood of oak, 1½ drachms
balsam of Peru, ½ drachm
oil of saffron by distillation, 6 drops.
Mix. Insert 2 or 3 drops warm with bombace into the ears.

The following fomentation against poor hearing after smallpox is well known:
Take of: Castoreum, myrrh,
colocynth, of each 1 drachm
asarabacca, 1 ounce.
Mix with a little bit of sanicle-seed and *saffron*. Make a fomentation with warm wine to be applied to the ears morning and evening with a tritorium, with immediate results. Horst, *Appendix de variolis et morbilis*, p. 312.

II Tinnitus of the Ears

The philosophers claim that all winds are dry and every creature moist, Giovanni Francesco Pico della Mirandola, *Astrologia*, book X, ch. 9.

In each of the ears with tinnitus you will find this: an annoying sound right inside the ears, originating because of exhalations, air, pituitous humours accumulated in the brain or lifted up from the lower parts, or by a flea driving into the cochlea and the eardrum.

What is tinnitus of the ears?

Before other remedies, Stocker in *Empirica*, p. 43, as well as Montagnana p. 336, recommend the following remedies from Rhazes:
Take of: White hellebore, *saffron*,
saltpetre, castoreum, of each 1 drachm
vinegar, strained honey, of each 1 ounce.
Boil together. Apply to the ear several times with anointed lint.

Or
Take of: Grubs found within the bark and wood of oak, 10 in number
washed woodlice, 15 in number
saffron, castoreum, of each 1 scruple.

Boil in oil of bitter almonds, 1 ounce, and vinegar, 2 ounces, to their consumption, press, and strain. Pour into the ear 1 or 2 drops of this morning and evening, and close the ear with cotton.

III Earache

According to Cicero, *Tusculanae quaestiones* II, pain is a disagreeable business. Undoubtedly harsh, hostile to nature, difficult to endure and to bear, as Seneca states very well in the *Consolatio ad Apollonium*, since many dispositions torment the mind.[2] The nature of pain is such that it makes it the most unpleasant thing of all; pain was for some even cause of madness, and of other incurable diseases, and some took their own life because of the intensity of pain.

What is earache?

Among all the other pains, earache is the fiercest; undoubtedly a disagreeable feeling caused by inflammation or by cold humours swelling out parts of the ears, or by a sore.

In cases of inflammation:

Take of: Thebaic opium dissolved in spirit of wine, ½ scruple
castoreum, 1 drachm
myrrh, 1 scruple
powdered saffron, ½ scruple.
Mix with milk, 3 ounces. Administer in an earthen vessel.

Let it be marked: Umschlag zum Ohr ["Poultice for the ear"].

Take of: Breadcrumbs or barley-meal, ½ pound
sape, as much as suffices.
Make a cataplasm, add:
opium, *saffron*, of each 2 drachms.
Mix. Pereda.

When the inflammation has a tendency to suppurate, administer the following remedy, which is truly successful:

Take of: Red onion, cooked under hot ashes, 1 in number
fresh butter, 2 ounces
oils of chamomile, roses, of each 1 ounce
saffron, 1 drachm.

2 Seneca wrote the *Consolatio ad Polybius*; the *Consolatio ad Apollonium* was written by Plutarch.

Mix. Make a cataplasm to be applied to the ear moderately warm and to be left in place day and night. Benedetto Vettori da Faenza; Rivière; Hendrik de Roy; Zecchio in *Consultationes*, p. 636.

Not differently, Grüling, *Medicinae practicae*, book I, part 3, p. 123, employs a similar remedy, which he claims to be miraculously beneficial, after taking pills and applying cupping-glasses, when the earache has a cold origin:
Take of: Red onion, cooked under ashes, 1 in number, add:
oil of chamomile, 1 ounce
dill,
fresh butter, of each ½ ounce
saffron, 1 scruple.
Make a plaster to be applied warm day and night. It takes away the pain, dissolves the abscess and it is good in cases of suppuration.

Or
Take of: Panic, 1 pugil[3]
wild marjoram, marjoram, of each 1 handful
flowers of chamomile, 1 pugil
saffron, ½ drachm.
Reduce to powder and make a little bag to be applied to the ear. Felix Platter, *Praxis medica*, p. 196.

When there is an abscess, Feynes, *Medicina practica*, book II, ch. 3, p. 34, prepares a liquor with the addition of *saffron* to clean the abscess.

IV Parotitis

Every pride, if it makes a ruler known for ambition, sumptuousness, too much retinue, or in any other way, is a path to ruin. In fact, as pride grows, fortune decreases (motto of Flavius Julius Constans). And Evagrius considers this true to such a degree that he defines pride as the inn of thieves.

When unruly humours accompanied by malignity hit the head, the main part of our body, and cause parotitis by lodging behind the ears, there is no-one who does see that often all the fortune of health is indeed very close to dangerous ruin.

3 Panic: a cereal, see Glossary.

What is parotitis?

Parotitis is a tumour of the glands around the ears with or without inflammation, caused by pituitous bile or bile mixed with blood, often by a malignity, or by crises during fevers.

The following liniment does wonders in softening and removing the tumour of the parotid glands.

Take of: Oil of henbane by expression, 1 drachm
of chamomile by distillation, ½ drachm
of saffron by distillation, 5 drops.
Mix.

Paré commends this plaster:
Take of: Plaster of oxycroceum,
of melilot, of each as much as suffices.
Soften with oil of chamomile.

V Ear Discharges

Pliny

One may not seek victory in every circumstance, especially if haste accompanies things not thought over. In fact, as Thucydides, book III, said, the things that hinder judgement are twofold: rage and haste.

We notice that very often ear discharges are ill-treated by some, especially in little babies and newborns, in such a manner that instead of the victory that they hope to gain for themselves with the cure, they hasten epilepsy or another affection of the brain, thus Hippocrates, book III, *Aphorism* 24. In fact, nature expels the excessive excretions in children mostly through this way, although it is a less suitable way.

How to take care of this: first the dirt is to be removed delicately with an ear pick and the ears are to be cleaned with a needle wrapped in very soft wool in a little bundle. Then, once the ears are cleaned, instil 3 drops of the following liquor in the leaking ear, or insert into the ear some cotton wool imbued with lukewarm liquor.

Take of: Finest honey, red wine, of each ½ ounce
alum, *saffron*, saltpetre, of each 1 scruple.
Grind and let them boil together a little. François Joël, *Practicae*, book IX, section 2, p. 505.

VI Ulcer of the Ear

Sharpness can penetrate everything. Thus, Hannibal, General of the Carthaginians, leading an army from Gallia to Italy across the impassable Alps, broke asunder a rock, which was the only possible route, by pouring vinegar into it. Thus he created himself a passage through the Alps. Livy, *Third decade*, book I.

The archeus of our body imitates Hannibal, as it often produces an ulcer in the ears with sharp and corrosive humours breaking through the cartilaginous parts of the ear. The ulcer of the ear is a relaxation with continuous discharge of pus, caused by consuming sharp and pungent humours, or by a burst abscess, or a blow, a sting, a fall or a wound, etc. in the middle ear or under the ears.

What is the ulcer of the ear?

To clean, dry and heal the ulcer, instil the following into the ear, by which one man's health was restored, see Rodrigo da Fonseca, *Consultationes medicae*, 39, p. 263:

Take of: One sweet pomegranate
boil in white wine, when it is boiled, pound in a mortar and extract the juice through sack-cloth, then:

Take of: This pomegranate-juice,
honey of roses, of each 1 ounce
turpentine, ½ ounce
aloes, frankincense, *saffron*, of each 2 scruples.
Mix. Make a remedy.

Or

Take of: Phlegmatic spirit of urine
of a young boy, 2½ ounces
aloes, frankincense, myrrh,
eastern saffron, of each ½ drachm
crocus metallorum, ½ scruple.
Let them stay in infusion overnight in a closed glass. In the morning make a colature, and instil some of it, lukewarm, into the ear by means of an auricular clyster. It wipes away and dries the ulcer, and it settles the flowing of humour.

CHAPTER XII[1]

Diseases of the Teeth and the Tongue

I Odontalgia

Diogenes Laërtius, book IV, ch. 7

Often one does not distinguish the truth from a lie, as one does not distinguish crystal from ice, if not for how long they last. While a lie easily melts like ice under the beaming rays of the sun of intelligence and careful enquiring, and all the steam disappears, the truth, instead, persists unchanged, sound and genuine forever, and protects itself easily from false plots. And whatever the appearance of one's own body is, that belonging to others is better. Observe that dental arch, I think that the teeth that now you would swear are like ivory (by all means like the teeth of the eastern Queen Zenobia, who was believed to have a mouth full not of teeth but of white pearls when she spoke and laughed), tomorrow you will realize are like coal, full of cavities and pain.[2] As they are below, the bones perceive the pain and, and thus certainly, like Job, they curse the day. Toothache is the abnormal suffering of the very spirit which vivifies the tooth, an abnormal affection caused by waste matter, or according to others, a relaxation caused by serous, bilious and pituitous humour, or by worms, or by air.

Gregory, *Moralia in Job*, book III

What is toothache?

To reduce the pain, the excellent Sebastian Georg Jung, PhD and MD, practitioner in Vienna, dearest friend and colleague, shared with me the recipe of this liquor in a letter:[3]

Take of: Shavings of wood of guaiacum, 2 ounces
sassafras, sarsaparilla, of each 1 ounce
root of pellitory of Spain,
raw alum, prepared saltpetre, of each ½ ounce
spagyrically-prepared opium, cloves, of each 1½ drachms
wild thyme, dittany of Crete,
saffron, of each 1 drachm.

Mix the ground ingredients in a glass, add:
spirit of wine and vinegar of wine, as much as suffices.

1 There is no Chapter XI in the original *Crocologia*. We have deliberately maintained this error to make it easier to compare chapters in the original text and our translation.

2 Septimia Zenobia, c. 240–c. 274 BC, queen of the Palmyrene Empire in Syria.

3 Again, Hertodt has Jung's Christian names in the wrong order.

 | DOI:10.1163/9789004435292_015

Extract a tincture; to be kept in the mouth, warm, for a long period.

Or

Take of: Long and white pepper, of each 3 grains
black pepper, 4 grains
saffron, 1 scruple.

Grind coarsely, boil together in white wine, 3 ounces, to the consumption of the third part, then add by dissolving:

Thebaic opium, 3 grains.

Let it boil again a little on a slow fire. Keep often in the mouth, warm. Finck, *Encheiridion dogmaticum*, p. 59.

Or

Take of: *Saffron threads*, as much as you please.

Let them burn up on a tin plate, collect the oil which comes out with bombace and place it on the aching tooth.

Additionally, Zecchio describes a cataplasm which soothes the pain with the core of excorticated sweet apples, *saffron*, etc. *Consultationes medicinales*, 36, p. 393.

Or

Take of: Core of bread, 3 ounces.

Boil in milk, add:
seed-meal of linseed,
oils of roses, chamomile, of each 1 ounce
saffron, 2 scruples.

Pound and apply to the affected area.

Or

Take of: Opium dissolved in vinegar,
powdered saffron, of each 1 scruple
choice mastic, ½ drachm
gum caranna, 2 scruples
gum elemi, resin, tragacanth, of each 2 drachms.

With balsam of mastic make a plaster to be applied on the jaw.

11 Prunella

Mock suns are never seen without fear, in fact they forebode either new covenants or arranged conspiracies or disorders, and are indicators of hidden plots.[4] The aforementioned was clearly confirmed by the mock suns seen this year, 1670, on February 12th in Hungary.

A great and serious disorder, and often a concealed plot, is thought to be going on in the body when prunella appears on the tongue, a very common sign of severe illnesses, as Galen puts it, *Commentaria 1* in *6 Epidemiarum, Circa principium de uteri phlegmone*; and Hippocrates, *De morbis*, III, when he speaks of a certain burning fever, "there is fever and ardent thirst, and the tongue is black and rough."

What is prunella?

Prunella is the blackness and roughness of the tongue, caused by fever or by some inflammation of the internal organs.

If it is caused by an inflammation of the liver, Rhumel applies the following epithem on the region of the liver:

Take of: Camphor, ½ drachm.
Dissolve in 1 spoonful of spirit of wine, add:
saffron threads, 7 in number
water of nightshade, 2 spoonfuls
vinegar of roses, 1 spoonful
spring water, half a pint.

Mix everything together, soak a blue woollen cloth in it, and apply repeating seven times, then rinse in running water, and you will find by experience that the remedy is effective.

Merit of the number seven

It appears that this author attributes great importance to the number seven, perhaps driven by the theory that the number seven, if the individual parts of it are considered, is full of various power, and every greatness. It is the number upon which the entire conception, forming, feeding and life are based.

It is the number of happiness:

… … … Oh! Three times and four times happy.

It is a number extremely suitable for purification, as in the *Book of Kings*, ch. 4, on the leper, "Go and wash yourself seven times in the Jordan, and your

4 A mock sun, sundog or parhelion: a bright spot on either or both sides of the sun, caused by light interacting with ice crystals in the atmosphere.

flesh shall be restored and you shall be cleansed", and a little further, "And he dipped himself seven times in the Jordan, and he was clean. He shall dip himself seven times in the blood of a sparrow and he will be cleansed from leprosy. He who suffers a flow of blood, shall wash for seven days with water."

And it is the number of rest: on the seventh day Moses stopped speaking; on the seventh day Our Lord the Creator abstained from work. According to Pythagoras it is the vehicle of human life. According to Aristotle, *De historia animalium*, 7, it is extraordinary, because the number seven most particularly marks changes.

He who wishes to know more, see André du Laurens, *De crisibus*, ch. 9; Fabrizio Padovani, *De anno climacterico*; among other writers, Galen's *De diebus decretoriis*, and others.[5]

In cases of prunella, others apply egg white with *saffron* and spirit of wine to the region of the heart.

III Paralysis of the Tongue

See Diseases of the Brain, § 2.

---

5 *De anno climacterico* is in fact by Giovanni Battista Selvatico, not Padovani. See Hertodt's reference list, in Appendix 1.

CHAPTER XIII

Diseases of the Chest

1 Angina

Sometimes it is difficult to predict the reasons why you are loved; as Ausonius wrote, happiness is often that which we cannot explain.

One may say the same about a vase made of ivy used in place of a beautifully wrought bowl, and preserving from angina.

What is angina?

Angina is the difficulty of swallowing and breathing because of the inflammation and tightness of the uppermost part of the throat, larynx and pharynx.[1] It is caused by burning blood or phlegm, melted by the heat and coagulated again in the pharynx because of the incoming cold, or, according to others, caused by acid and lixivial salt diminishing with continuous fever.

A gargle that wards off and cures angina is such:

Take of: Roots of liquorice, 1 ounce
marshmallow, ½ ounce
leaves of the medium consound,
flowers of self-heal, of each 2 handfuls
dates, fat dried figs, of each 4 pugils
seeds of marshmallow, fenugreek, of each 2 drachms.

Boil in a sufficient quantity of strained water. Take of: 1½ pounds.

Add:
honey of roses, 2 ounces
ashes of swallows, 3 drachms or ½ ounce
saffron, ½ scruple
camphor, 2 or 3 grains.

Make a gargle. Rolfinck.

When it gets worse:

Take of: Fat figs, 6 in number
dates, 3 in number

1 Hertodt's definition of angina is quite different to today's, viz. a chest pain caused by reduced blood flow to the cardiac muscles.

 | DOI:10.1163/9789004435292_016

balaustines, ½ ounce
seeds of linseed, fenugreek, of each ½ ounce.

Boil in water, 2 pounds, down to half, strain and in the strained broth add:

diamoron,
strained honey of roses, of each 1½ ounces
swallows' nests dried and powdered, 2 drachms
saffron, ½ drachm
woman's milk, 1 ounce.

Make a gargle to be used frequently, lukewarm. François Joël, *Practicae* II, section 8, p. 358.

Mynsicht's cataplasm of swallows' nest (to which, not without cause, the author added *saffron*) smeared often on the neck is a panacea against irremediable angina.

If the swelling does not go away and tends to suppurate, since it usually lasts for four or five days, one will need to accelerate its course, which the following aromatic wine does in an excellent way:

Take of: Extract of pith of fresh cassia, 1 ounce
magistery of tobacco,
saffron, of each ½ drachm.

Keep some of it in the mouth, frequently, so that in melting, it goes down the throat little by little.

Additionally, for accelerating its course:

Take of: Core of a wheaten loaf, ½ pound
root of white lilies, marshmallow, of each 1 ounce
seeds of linseed, 6 drachms; fenugreek, ½ ounce.

Boil in sweet milk, as much as suffices, and add:

oil of sweet almonds,
white lilies, of each ½ ounce
fresh unsalted butter, 6 drachms
saffron, ½ drachm
egg yolk, 1 in number.

Make a cataplasm. Balthasar Timaeus von Güldenklee, *Casus medicinales*, book II, ch. 1, p. 78.

Or

Take of: Swallows' nest, 1 in number

Boil down in a decoction of marshmallow to the consistency of a pottage, add:

oils of sweet almonds, dill, of each 2 ounces
saffron, 1 drachm
cassia, 1½ ounces.
Make a cataplasm.

When the coagulated phlegm becomes harmful, one must refrain from medications that thrust it back, so that the humour causing the inflammation, removed with such remedy, does not go back to the principal part and does not obstruct the passage of air, resulting in fatal ruin.

This ointment is serviceable:

Take of: Oil of wallflower, or orris, 2 ounces
ointment of marshmallow or the restorative ointment, 4 ounces
saffron, 2 scruples.
Mix and make an ointment. Anoint the neck and the region of the throat with it. Feynes, *Medicina practica*, book III, ch. 1, p. 315.

If it is degenerating further:
Take of: Fresh butter or hens' grease, 2 ounces
seed-mucilages of linseed, fenugreek, of each 1 ounce
powder of swallows, 1 ounce
or swallows' or doves' nests, 1½ ounces
orris, *saffron*, of each 2 drachms
oil of lilies, as much as suffices.
Mix. Idem, op. cit., or:

Take of: Ashes of swallows, 3 ounces
dried flowers of chamomile, ½ drachm
saffron, spike, of each 1 drachm.
Grind and add to honey, and rub over the region of the throat. Idem, op. cit.

II Asthma

Common is the saying:

Vetus asthma
Mirabile Phantasma
Quod nullum curat cataplasma

["Old asthma
strange phantom
that no cataplasm cures"]

In the same way as asthma is difficult to cure in young people, thus it is incurable in the elderly. But he who can hope for nothing, despair of nothing, because *saffron* will comfort even the desperate.

What is asthma?

Asthma is the difficulty of breathing because of the tightness of the respiratory apparatus, caused by thick and viscous humours deposited in the lungs while circulating, or by a small swelling or by stones, etc.

In this case, phlegmagogues are administered uninterruptedly mixed with remedies for the chest, as is the following nodulus, frequently used also in cases of danger of suffocation:

Take of: Herb of horehound, hyssop,
maidenhair, sundew, of each 3 pugils
flowers of elecampane, 1 pugil
saffron, ½ scruple
root of orris, flowers of elecampane, of each 6 drachms
polypody of oak trees, 1 ounce
whitest mechoacan, 6 drachms
seed of safflower, ½ ounce
whitest agaric, 3 drachms
seed of nettle, 2 drachms
cream of tartar, 2 drachms
currants, 2 ounces.

Make a nodulus.

Let it be marked: Brustpischl ["Chest ?nodulus"][2]

From the manuscript of the excellent Dr Johann Michael about Jonston.

Take of: Whitest agaric, 2 drachms
mechoacan, 1 drachm
seeds of field *saffron* or safflower,
fennel, anise, of each 1½ drachms
Austrian saffron, 6 grains.

Besprinkle with 7 drops of the *asthmatic spirit of gum ammoniac and saffron* described below, then pour on wine of elecampane, as much as suffices. In the morning strain and filter for a dose.

2 *"Brustpischl"* could not be fully translated and may be a word that was typeset incorrectly.

Or

Take of: Mass of pills of gum ammoniac, 1 scruple
extract of spurge, 8 grains
saffron, 3 grains.

With *pectoral elixir of saffron*, make into 13 pills for a dose.

Or

Take of: Dissolved magistery of turpeth with pine-nuts,
agaric, of each ½ scruple
saffron, 4 grains.

Mix and make a powder to be taken in wine or hydromel.

We have noticed that a decoction of an old cockerel prepared in the following way with *saffron* is often very beneficial:

Take of: Roots of liquorice, elecampane, orris, of each ½ ounce
herb of horehound, leaves of hyssop, coltsfoot,
scabious, of each 1 handful,
flowers of elecampane,
field *saffron* or field safflower, 1½ pugils
Austrian saffron, ½ drachm
raisins, figs, of each 6 drachms
jujubes, 12 in number
troches of agaric, ½ ounce
white mechoacan, 3 drachms
turpeth, 2 drachms
ginger, 1 drachm.

Boil with an old cockerel in clean water and when the broth is freed from the fat, administer the draught in the morning. If this does not move the bowels enough, add some Calabrian manna.

With paroxysm, a vomit is often miraculously good, because it expels not only what clings to the stomach and the digestive tract, but also what is in the bronchial tubes of the lungs:

Take of: Ruland's aqua benedicta,
tobacco, of each ½ ounce
oxymel of saffron, 3 drachms.

Make a draught to be taken lukewarm.

Additionally, the following are good for cleansing and for the expectoration of thick humours:

Take of: Lohoch sanum, 2 ounces
pour into:
hot spring water, 1 measure.
Let them stay during the night, lukewarm, strain with a linen cloth and add:
dried turnip, 1 handful
figs, 1½ ounces
seed of anise, ½ ounce
saffron, 1½ drachms.
Boil for half an hour, press with a linen cloth, then:
Take of: The colature, 1 pound
red sugar candy, ½ pound.
Let them boil and make a syrup.

Or
Take of: Roots of liquorice, 1 ounce
elecampane, iris, of each ½ ounce
maidenhair, hyssop, of each 1 handful
pitted raisins, figs, of each 1 ounce
saffron, 1 scruple.
Boil, add:
honey, as much as suffices.
Drink in cases of difficulty of breathing. Felix Platter, *Praxis medica*, p. 390.

Or
Take of: Zedoary, ½ ounce
gum ammoniac, 2 drachms
flowers of sulphur, 3 drachms
saffron, 1 scruple.
Boil down to half in hydromel.

Let it be marked: **Lufft-tranck wohl umzurüttlen/zwen/dren** ["Breath-drink, to be shaken well, two/three"] spoonfuls.

Or
Take of: *Saffron*, ½ scruple
musk, 1 grain
clarified juice of basil, 2 drachms
white aromatic wine, warmed, 4 ounces.
Make a draught to be administered warm. Zecchio, *Consultationes medicinales*, p. 580, no. 53.

The following is an arcanum against this affection:
Take of: *Saffron*, ½ scruple
musk, 3 grains.
Mix with decoction of iris root.

Or instead of this:
Take of: *Moschardini of saffron*, ½ scruple
To be taken with a decoction of iris root.

Take of: Waters of scabious, elderflowers,
blessed thistle distilled with wine, of each 2 ounces
extracts of iris, scabious,
eastern saffron, of each 6 grains
oleosaccharum of anise, ½ ounce
syrup of hyssop, as much as suffices according to taste. Werner Rolfinck, *Methodus medicinae specialis*, book VIII, ch. 10.

The poor can benefit from a decoction of hyssop besprinkled with *saffron*.

Or
Take of: 5 or 7 raisins.
Divide lengthways, cover again by laying on top some *saffron* threads and, in a bowl, besprinkle with sugar candy. Pour over spirit of wine, which will be burned up. Eat the aforementioned raisins in the morning, drinking afterwards the spirit of wine left after the burning in the form of a syrup, and repeat this every morning and evening if you wish.

Joël has experimented with the following draught against asthma:
Take of: Wine of woodlice, 8 ounces[3]
syrups of hyssop, horehound, of each 2 ounces
essence of elecampane, 2 drachms
saffron, ½ drachm.
Mix. Make a draught for the doses.

Against the asthma of children we pour wine with *saffron* and cinnamon over millipedes and we sweeten it with a little bit of white sugar candy for a very successful result.

3 The original line here is "*vini millepedum vel asellor.*" and it appears that Hertodt is using *millepedum* and *asellorum* as synonyms, rather than these being two different wines.

Mynsicht's *spirit of juniper and saffron* is extremely good for asthma, and so is the following spirit, which is an arcanum. Personal communication of the excellent Dr Matthias Franz Hertodt, medical practitioner, a very successful practitioner, by me never praised enough:

Take of: Verdigris, 1½ pounds
gum ammoniac, 1 pound

Then distil a spirit and, once it has separated from the oil, rectify it, 1 drachm of *Austrian saffron* having been placed on the beak of the alembic. Store the dyed spirit for use.

Arcanum against asthma namely *asthmatic spirit of gum ammoniac and saffron*

Let it be marked: *asthmatic spirit of gum ammoniac and saffron*. Dose: 10 drops to 1 scruple to be taken in *hydromel of saffron* or in another suitable liquor, like asthmatic water, water of lohoch sanum, or pectoral water, etc.

Eclegme or lohoch.

Take of: Ordinary lohoch for asthma,
lohoch of cabbage-stalk, of each 1 ounce
our round tablets of saffron, 1 drachm
powdered root of cuckoo-pint prepared with wine, 1 drachm [unclear]
prepared crab's-eyes, 1 scruple.

With *oxymel of saffron*, as much as suffices, make lohochs.

Let it be marked: Ablösch-safft ["Extinguishing juice"].

Take of: Gum ammoniac, 1 scruple; *saffron*,
sulphur pulverised very frequently in violet-water,
of each 5 grains
musk, ½ grain.

Mix with oxymel, as much as suffices, and make a soft bolus according to Art. Zecchio, *Consultationes medicinales*, 45, p. 485.

Our Phosphorus, the excellent Dr Philip Jakob Sachs von Löwenheim, PhD and MD, glory of our beloved College, made known to me in a recent letter that an apple, hollowed out and refilled with frankincense and *saffron*, roasted and eaten, is proven to be good for asthmatic subjects.

Take of: Extract of elecampane, 1 drachm
liquorice, 2 scruples
magistery of gum Benjamin, ½ drachm
extract of saffron, 6 grains.

Mix. Make into large pills.

Let it be marked: sonderliche Pillen gegen den Dampf in Mund zu nehmen ["Special pills against the vapour, to take by mouth"].

Pierre Potier, *Observationum et curationum, centuria* 1, ch. 77, p. 68, successfully cured a 30-year-old man, who seemed to be about to die at any moment because of asthma in the chest full of catarrh, by first administering extract of spurge three times, which drives the thick and sticky matter out of the lungs, then with the pills about to be described:

Take of: Socotrine aloes, ½ ounce
myrrh, gum ammoniac, of each ½ drachm
eastern saffron, 1 scruple
flowers of sulphur sublimated three times with its vitriol, 2½ drachms.

With syrup of white horehound juice, make a mass.

Dose: 1 ounce for several days.

Take of: Choice myrrh, frankincense, of each 1 drachm
tragacanth, gum Arabic, of each 3 drachms
saffron, liquorice juice, of each 1 drachm
flesh of 5 dates.

With sape, make sublingual pills. Brunner, *Consilia,* p. 181, no. 33.

Take of: Flowers of sulphur sublimated on its own, 1 scruple & 2 grains
gum Benjamin, 5 grains
saffron, 2 grains
musk, 1 grain

sugar to the equivalent weight of all the ingredients. Mix. Make a powder. Dose: the point of a knife in a spoonful of wine. Rolfinck.

Or

Take of: Faeculae of cuckoo-pint, 1½ drachms
magistery of gum Benjamin,
saffron, of each 1 drachm
sugar, 2 drachms
oil of anise by distillation, ½ drachm. Mix.

Let it be marked: Lufft- und Brust-Pulver ["Breath and chest powder"].

Take of: Gum ammoniac, 4 grains
saffron, 3 grains
species diamoschi dulcis, 6 grains.

Mix. Make a powder.

Or the following, which is of visible help in cases of difficulty of breathing:
Take of: Fresh spermaceti, 2 scruples
oleosaccharum of saffron, 7 grains.
Mix. For a dose to be administered in warm beer.

Powder for meals:
Take of: Foxes' lungs, prepared according to Art, 1½ drachms
liquorice, 3 drachms
powder of dried hyssop, 2 drachms
orris, anise, of each 1½ drachms
saffron, ½ drachm
whitest sugar, 1 ounce. Make into a powder. Dose: 2 or 3 drachms.

To be sprinkled upon food or used in broth. Feynes, book III, ch. 5, p. 353.

To this, let external remedies be added, with which it is no secret that the affected parts are relieved:
Take of: Seed-mucilages of fenugreek, linseed, marshmallow, extracted with marshmallow-water, 1 ounce
leaves of coltsfoot, ½ ounce
unsalted butter, ½ ounce
fresh hens' grease, 2 ounces
powdered orris, 3 drachms
saffron, ½ drachm
oil of sweet almonds, 5 ounces.
White wax, as much as suffices. Make an ointment. With this anoint the chest before going to bed, rubbing well, and in the morning wipe off with a linen cloth. Johann Wittich, *Consilia medicinalia*, p. 244.

Take of: *Oil of saffron*,
white lilies, of each 1 ounce
powders of iris, *saffron*, round birthwort,
opopanax, of each 1 scruple.
Rub everything together to pieces extremely finely, and then make a warm ointment. With this anoint the chest. Bartolomeo Montagnana, *Consilia*, no. 77, p. 404. And also, *Consilia*, no. 33, *De aegritudinibus cerebri*, p. 218, demonstrates that *saffron* is very useful against breathing problems.

Take of: Oils of white lilies, sweet almonds,
violets, of each 2 drachms
capons' fat,
ointment of marshmallow, of each ½ drachm
saffron, 2 scruples
wax, as much as suffices. Make an ointment. Rolfinck, *Methodus medicinae specialis*, book VII, section 2, ch. XI.

The ointment with an onion hollowed out and refilled with *saffron* and frankincense, listed in Philip Jakob Sachs, *De usu praeparationis croci*, does good service too.

Or

Take of: Hens' fat, ½ pound
deer marrow, 3 drachms
ointment of marshmallow, 5 drachms
oils of white lilies, wallflower,
sweet almonds, of each ½ ounce
powders of iris, elecampane, of each 2 scruples
saffron, 1 scruple.
With wax, make a liniment. Balthasar Timaeus, *Casus medicinales*, book II, p. 96.

Or

Take of: Oils of white lilies,
sweet almonds, of each 2 ounces
fresh butter, fat of hens, geese, of each 1 ounce
seed-mucilages of linseed, fenugreek
extracted with coltsfoot-water, of each 6 drachms
dried hyssop, iris, of each ½ scruples
saffron, 1 scruple.
With wax, make a liniment. Zecchio, *Consultationes medicinales*, p. 53.

Or

Take of: Oil of wallflower, 2 ounces
bay, marshmallow, of each ½ ounce
seed-mucilages of linseed, fenugreek, of each ½ ounce
fresh butter or calf marrow, 3 drachms
calamite storax, 1½ drachms
powders of iris, round birthwort, of each 1 drachm

saffron, ½ drachm
wax, a little.

Make an ointment. With it, warm, smear the chest, rubbing well.

Take of: Plaster of oxycroceum,
gum ammoniac, as much as suffices.

Soften with *oil of saffron by insolation*, place on a patch, administer in the customary form.

Additionally, Hoefer in *Hercules medicus*, book II, ch. 2, p. 104, reports of a certain epithem of Thomas Mermann, once the Galen of Bavaria, in which *saffron* is one of the ingredients too.

When the violence of the disease is so great that we are forced to begin the cure with laudanum, Pierre Potier, *Observationum et curationum, centuria* 3, p. 228, uses the following pills with *saffron*:

Take of: *Eastern saffron*, 3 grains
Thebaic opium, 4 grains
musk, 1 grain.

Make a pill with a little drop of some liquor. Take one hour after eating dinner.

III Coughs

A cough is like a hidden snare, against which many dash, while, exceedingly unconcerned, they are troubled by no disease.

A cough is a hidden snare, as it, being frequent and long-lasting, very often leads to the spitting of blood, as Avicenna testifies. Sometimes, when it is violent, it suffocates. Often, when strong, since the chest and the lungs are heated, it causes fever. It is dangerous in elderly people and accompanies until death. Hippocrates, 2, *Aphorism* 40. It is bad, in fact irremedi able, if it affects a dropsical person; 6, *Aphorism* 35 or section 7, *Aphorism* 47.

And also, if it causes a hunch before puberty, they regularly die; 6, *Aphorism* 46.

What is a cough?

Coughing is the distorted motion of the lungs, with which the irritated excretive capacity tries to expel, with a noise, that which troubles the lungs. It is caused principally by catarrh or by pungent sharp particles irritating the bronchial tubes of the lungs, or by thick particles, accumulated while circulating, stuffing the air passages of the lungs.

The lungs are cleared from these thick particles by those remedies that possess the power to cut through and to expel, and that in a certain way, soothe and moisten, which have been indicated in the previous chapter for asthma, as well as by a distilled spirit against a cough (François Joël, *Practicae*, book I, p. 268, to which *saffron* is added), or the following julep:

Take of: The *compound sugared oil of saffron*, 1½ ounces
syrups of scabious, elecampane,
coltsfoot, of each 1 ounce
waters of coltsfoot, scabious, of each 1½ ounces.
Mix. Dose: 2 or 3 spoonfuls, often.

Or
Take of: The *pectoral elixir with saffron*, ½ ounce
the *asthmatic spirit of gum ammoniac and saffron*, 2 drachms.
Mix. Dose: 30 or 40 drops in broth of lean meat.

Take of: Shavings of liquorice root, elecampane, of each 1 drachm
leaves of hyssop, ½ drachm
blessed thistle, 2 drachms
prepared foxes' lungs, 1½ drachms
saffron, ½ drachm
Manus Christi sugar to half of the equivalent weight of the ingredients. Make a powder. Samuel Schönborn, *Manuale medicinae practicae*, p. 67; Christoph Petri, *Asylum languentium*, p. 117.

Against children's coughs, Philip Grüling describes an excellent powder, which *saffron* improves, see *Praxis medica*, book II, part 2, p. 176.

Or
Take of: Root of orris, 1 ounce
spermaceti, 2 scruples
myrrh, 1 drachm
saffron, 14 grains
sugar candy, 2 ounces.
Make a powder to be administered in beer several times.

When a cough is caused by catarrh:
Take of: Species diaireos, 1½ drachms
flowers of sulphur, 2 scruples
extract of elecampane, 1 scruple

saffron, 8 grains
prepared amber, 1½ drachms.

With a sufficient quantity of sugar dissolved in asthmatic water, make round tablets according to Art. Alternatively:

Take of: Species diaireos, 4 scruples
faeculae of cuckoo-pint, 1 drachm
extract of elecampane, 1 scruple
saffron, 6 grains

with a sufficient quantity of sugar dissolved in coltsfoot-water, make round tablets according to Art. Besprinkle with oil of anise.

Let it be marked: Brustzelt ["Chest troches"].

Take of: Species diambra, hyssop,
liquorice, gum Arabic,
saffron, of each 1 scruple

syrup of hyssop, of liquorice, of each 1 ounce; penides, 6 drachms. Incorporate all the ingredients together and administer in an earthenware container as an electuary. Johann Wittich, *Consilia medicinalia*, p. 126.

A diasulphur is useful against any old cough, and for the cleansing of the chest:

Take of: Powder of yellow sulphur, 1 pound
cinnamon, 3 drachms
saffron, 2 scruples
ginger, 2 drachms
musk dissolved in rose-water,
if the patient pleases, 8 grains

white honey, as much as suffices, Make an electuary. Dose: ½ ounce per time in the morning and in the evening, and then make tablets to keep in the mouth. Fonseca, *Consultationes medicae*, p. 499.

The following is a panacea against all kinds of cough:

Take of: The withered lung with the gullet of a deer.

Then make a powder. To 1 ounce of this powder add:
magistery of saffron, ½ drachm.

With the best honey make an eclegme to be used every day.

When, the lungs having been cleansed, the cough troubles very much in the course of the cure and causes wakefulness, it should be suppressed in inverted order with globules of storax.

For example:

Take of: Liquid storax, frankincense,
mastic, myrrh,
opium, *saffron*,
juice of liquorice, of each [no quantity given].

Make pills of the size of a chickpea; administer 1 or 2 maximum. Feynes, *Medicina practica*, book III, ch. 4, p. 346.

Additionally:

Take of: Washed butter, 4 ounces
oil of violets, 1 ounce
sweet almonds, 2 ounces
saffron, ½ drachm
white wax, a little.

Mix. Make an ointment, incorporate together. To this you may add: hen's fat, ½ ounce. Mix. Stocker, *Empirica*, p. 64.

Or

Take of: Oils of lilies, orris, of each 2 ounces
dried root of elecampane, ground, 3 drachms
dried leaves of coltsfoot, 2 drachms
dried hyssop, 1½ drachms
saffron, 2 scruples
hens' fat, ½ ounce
lanolin, 1½ ounces
wax, as much as suffices.

Make an ointment which is to be smeared, warm, on the region of the chest. Feynes p. 347.

IV Phthisis

God usually reveals the next changes that are going to happen in the affairs of states mostly with signs or portents, and in other ways. Thus, during the night when Troy was captured, some gods were seen (as Virgil says) leaving the temples and carrying out their images.

Among the various signs which indicate that phthisis is already verging towards its terminal stage, and that the following change is going to be death, is diarrhoea: thus, Hippocrates, book V, *Aphorism* 12, "in persons affected with consumption, the hair falls; they die when diarrhoea occurs."

Phthisis is the consumption of the whole body because of an ulcer in the lungs. It is caused by the filling up and breaking of circular vesicles due to very thick and sharp water, or due to other things corroding or damaging the lung itself. What is phthisis?

With this affection, Kneuffelius' anti-phthisic round tablets do wonders:

Take of: Flowers of sulphur thrice sublimated, 1½ drachms
diaphoretic antimony well calcined with three times
as much of saltpetre so that all the emetic power (or rather Potier's
anti-hectic) is taken away, ½ ounce
mastic, whitest amber, of each 1 drachm
species diatragacanthi frigidi, 2 drachms
saffron, ½ scruple
oil of anise, 6 drops
sugar dissolved in water of ground-ivy, 3½ ounces.
Make large round tablets.

Or
Take of: Powdered flowers of sulphur,
liquorice shavings, of each 1 drachm
root of orris, 2 scruples
Haly's anti-phthisic powder, 3 drachms
gum Benjamin, *saffron*, of each ½ scruple
whitest sugar, 5 ounces.
With rose-water, as much as suffices, make a paste, from which make troches.
Diemerbroeck.

For other troches, agglutinant and soothing the cough, of which *saffron* is an ingredient too, see Feynes, *Practica*, p. 371.
Take of: Sublimated flowers of sulphur with aloes and myrrh,
1 drachm
gum Benjamin, 1 scruple
leaves of sundew, 3 pugils
seeds of dead nettle, 1 drachm
roots of elecampane, orris, of each ½ scruple
gum ammoniac, myrrh, aloes, of each 8 grains
saffron, 10 grains; musk, 3 grains.
Mix.

Let it be marked: Lungen Triesenet [“Lung Powder”].

Take of: Flowers of sulphur, 1 pound
myrrh, 3 ounces; aloes, 2 ounces
saffron, 2 drachms.

Grind and pour over them spirit of turpentine to the height of two digits. Let it digest for 12 days in sand, then extract the essence of sulphur and of all the other dry ingredients. Filter, and you will have a balsam which is good not only against phthisis but also against pleurisy, the stone, fevers, plague and dropsy. Dose: from 2 drops to ½ scruple in an appropriate vehicle.

Ultimate balsam against the phthisis

Another ultimate balsam, passed on to me by the excellent Dr Matthias Franz Hertodt, medical practitioner, is this:

Take of: Milk of a woman breast-feeding a baby boy, 2½ pounds
core of wheat bread, 2 ounces
pot barley, bruised, 1 drachm
sublimated flowers of sulphur, ½ ounce
best aloes, myrrh, frankincense,
true mummy, of each 1 drachm
Austrian saffron, 2 drachms
seeds of dead nettle,
lungs of deer, foxes, of each ½ ounce
prepared crab's-eyes, 2 drachms
prepared red coral, 1 drachm
prepared pearls, 2 scruples & 5 grains
species de gemmis frigidi,
species diatragacanthi frigidi,
Haly's anti-phthisic powder,
prepared hartshorn, of each 1 drachm
Melitean earth,
dissoluble sugar of lead, of each 4 scruples.

Leave in infusion for 3 days and 3 nights and distil with a glass retort for a balsam. Rectify it with rose-water in which root of orris has been infused, and store for use.

Let it be marked: Anti-phthisic balsam for wounds.

To the magma remaining from the distillation add:

Take of: Leaves of mouse-ear hawkweed, 10 handfuls
herb of ground-ivy, sundew,
red-flowered yarrow,
white sanicle,
coltsfoot, of each 2 handfuls

root of comfrey,
tormentil, of each 1 pound.
Make an anti-phthisic salt according to Art.

Well-known remedies with sulphur, because they dry up and consume, are not always beneficial if they are not mixed with humectants.

Christoph Wirsung, part 2, ch. 5, and, after him, our colleague Christoph Petri in his *Asylum languentium*, p. 115, prescribe water of capon improved with *saffron*, especially against the phthisis of infants.

Additionally:

Take of: Oils of wallflower, sweet almonds, of each 1½ ounces
seed-mucilage of linseed, 6 drachms
woman's milk,
sublimated snail-water, of each 1 ounce
saffron, 2 scruples
white wax, as much as suffices.
Make an ointment. Feynes, book III, ch. 7, p. 372.

For another pectoral ointment with *saffron*, remarkable when pain troubles the patient, see Johann Wittich, p. 264.

Take of: Seed-mucilage of quince, 2 ounces
fresh butter, 1½ ounces
powder of orris, 3 drachms
dried leaves of coltsfoot, 2 drachms
milk of a woman who has given birth to a baby girl, 1 ounce
oil of violets, 3 ounces; *saffron*, 1 scruple
white wax, a little.
Make a liniment to be smeared on the chest.

V Pleurisy

This whole life is nonsense and thorns. But when is this truer than in case of pleurisy? In fact, the latter, like a piercing thorn, tortures with severe and uninterrupted torment, and, "a suppressed pain suffocates and the heart burns inside," Ovid, *Tristia*, V. 1.

What is pleurisy?

Pleurisy is the inflammation of the pleural membrane accompanied by acute constant fever, a pricking pain in the side, and an annoying cough. It is caused by extravasated or stagnating hot blood, when the subtle spirituous particles tempering the blood fly away, while the sharp, sour and salty particles remain.

An excellent remedy is the powder of Dr Marcus Banzer, Dean of the Faculty of Medicine in Wittenberg, which contains *saffron*, see Dr Sachs von Löwenheim, *Gammarologia*, p. 664.

Additionally, Marcello Cumano, in Welsch's *Sylloge curationum et observationum*, proposes the following magistral ointment, which, premised a venesection, relieves and acts as a painkiller. With it the afflicted side is alleviated, the pain is mitigated, expectoration is increased, respiration is freer, and finally, the skin having been made imperceptibly thin, in some measure the pleurisy dissipates, and this helps with exhalation. When, with this condition, there is no opportunity to use repellents, or even at the beginning:

Take of: Oils of sweet almonds, sesame,
violets, of each 4 ounces
hens' fat, fresh butter, of each 2 ounces
white wax, 3 ounces
saffron, 2 scruples.
Mix.

Or
Take of: fat of ducks, hens, of each 1½ ounces
fresh butter, 1 ounce
saffron, 1 drachm
oil of violets, 2 ounces; white wax, as much as suffices.
Make an ointment. Stocker from Bartolomeo Montagnana.

Or
Take of: Fresh restorative ointment, 4 ounces
oils of violets, sweet almonds, of each 1 ounce
saffron, 1 scruple. Mix.
Feynes, *Practica*, book III, ch. 8, p. 386.

Or
Take of: Ointment of marshmallow, 6 drachms
oil of saffron by insolation,
of chamomile, Mynsicht's anodyne
oil of 7 flowers, of each 2 drachms[4]

4 The original text here was "*ol. Croci per insolat. chamomill. anodyn. 7. flor. Myns. ana ʒij*". We translated the latter part as "Mynsicht's anodyne oil of 7 flowers" but have not been able to find the identity of the seven flowers. This ingredient abbreviation also occurs in another of Hertodt's works, see his *Tartaro-Mastix Moraviae* (1699), p. 231.

deer marrow, 3 drachms
fat of mountain-mouse, 1½ drachms
extract of saffron, 13 grains

a little wax. Make a liniment for the anointment of the aching side. Administer by placing a lead-plaster on top.

Wolfgang Gabelkover, *Curationum et observantionum medicinalium, centuria* 1, third remedy, p. 5, prescribes a fomentation with a quartarius of goat's milk or cow's milk heated in a pan, adding two spoonfuls of clarified honey and *saffron*, which should dye the milk well. Apply the milk, poured sufficiently warm into a beef bladder, to the aching side.

In cases of spurious pleurisy, the same author prescribes another remedy to dissipate the matter accumulated in the outer muscles of the side:

Take of: Flowers of chamomile,
melilot, of each 1 handful
seeds of linseed, fenugreek, of each 1 ounce
bran, 2 ounces; *saffron*, ½ drachm.

Let them boil in a sufficient quantity of spring water until one third of the water is consumed, and apply the decoction in a beef bladder to the aching side. After this decoction has been warmed and applied a first and a second time, the pain decreases to the point that the patient can sleep well the whole night. *Centuria* 1, *curationum*, p. 2.

VI Pneumonia

Tyrants perform tragedies with a countenance pertaining to comedy, and, consequently, they are to be feared especially when they put on the fox skin.

The physician has something to fear when the severe pain in the side seems to be decreasing and soothing itself, but fever and difficulty of breathing increase, thus Hippocrates, section 7, *Aphorism* 11, "Pneumonia coming from pleurisy is bad." Pneumonia is the inflammation of the lungs with acute continuous fever, difficulty of breathing, coughing, and expectoration with traces of blood and an irksome feeling of heaviness on the chest. It is caused by hot blood spread in the tissues of the lungs, or in some measure stagnating.

What is pneumonia?

With this affection, venesection is the best remedy.

If the pneumonia is spurious, Feynes, ch. 9, book III, p. 393, prescribes this ointment:

Take of: Clarified juice of horehound, 2½ ounces
oil of orris, 1 ounce
fresh butter, 1½ ounces.

Let them boil together to the consumption of the liquid, and when they have been strained, add:

turpentine, pine resin, of each ½ ounce
powdered root of iris, 3 drachms
dried root of coltsfoot, 2 drachms
dry leaves of hyssop, 1½ drachms
saffron, 1½ drachms.

Mix. Make an ointment. Smear on the whole region of the chest.

VII Empyema

Vitriol found in mines destroys a miner's hope of further pursuing veins of metal. The somewhat bloody, feculent and fetid pus of the person affected with empyema imitates vitriol, for it takes away every hope of the doctor practicing an incision and furthering the cure, like said in Hippocrates, *Aphorism* 44, section 7, "when patients affected by empyema are treated by cautery or by incision, if pure and white pus flows out from the incision, they recover; but if the pus is somewhat bloody, dirty and ill-smelling, they die."

What is empyema?

Empyema is an accumulation of pus in the thoracic cavity, caused by advanced pleurisy, a wound, a blow, a fall, etc.

For the expectoration of pus:

Take of: Root of iris, previously boiled and bruised, 2 ounces
whitest sugar dissolved in hyssop-water, 1 pound.

Boil together, add:

cleaned sweet almonds infused in barley-water, 1 ounce
pine kernels, in the same way infused, ½ ounce
dried foxes' lungs, meal of bitter-vetch, of each 2 drachms
seeds of nettle, dried hyssop, of each 1½ drachms
powered liquorice juice,
ground seed of fennel, of each 1 drachm
saffron, ½ drachm.

Mix. Make a pandaleon. Hold frequently in the mouth. Feynes, book III, ch. 10, p. 395.

CHAPTER XIV

Diseases of the Heart

The Trojan horse deceived because it feigned an image for Minerva. Syncope often feigns a pleasant sleep, and deceives the observer to the point that the person whom you believe is sleeping a natural sleep, is already seized by a fatal sleep and sleeps through all eternity. Diogenes Laërtius

Hence, not wrongly, sleep is said to be the image of death and her brother.

Syncope is an abrupt and excessive failing of forces, because of a lack of heat and spirits with cold sweat, caused by pain, severe disease, poison, a malignity, menstrual retention, a wound, venesection, or restrained blood circulation. What is syncope?

In this case, the water of hart's-heart of the illustrious Dr Engelhard, Chief Physician in Anhalt, is very celebrated, as well as the water of hart's-heart of the illustrious Dr Behr of Strasbourg, remarkable because of *saffron*. For their recipes see *Elaphographia* by our very deserving colleague and our Pegasus, Dr Graba, p. 227 et seq.

Or

Take of: Water of tender hartshorn, 2 ounces
balm with wine, 1 ounce
cinnamon, borage, carbuncle stone, of each ½ ounce
arcanum or quintessence of saffron, ½ scruple
syrup of gems, 3 drachms. Mix.

Let it be marked: Röstliches Herz-wasser ["Reddish water for the heart"] when the cause is cold.

Or

Take of: Potent Rhenish wine, 2 ounces
waters of cinnamon, borage,
Mattioli's elixir of life, of each ½ ounce
essence of amber,
quintessence of saffron, of each ½ scruple.

Mix.

Take of: Waters of tender hartshorn,
scorzonera, Spanish round-leaved sorrel,
borage, the cooling water of Saxony, of each 1 ounce
quintessence of saffron, ½ scruple

 | DOI:10.1163/9789004435292_017

spirits of raspberries, strawberries,
black cherries, of each 5 drops
syrup of whole citron, 3 drachms.

Mix.

Let it be marked: Herzwasser ["Water for the heart"] when the cause is warm.

For a cordial powder with *saffron,* capable of rekindling inactive warmth, see Feynes, *Medicina practica,* book III, ch. 2, p. 324, and for a cordial powder of *saffron,* see Werner Rolfinck, *Methodus medicinae,* p. 703.

When the syncope is caused by menstrual retention, *essence of saffron* helps magnificently.

Additionally:

Take of: Rose-water, 3 ounces
sweet wine, 2 ounces; vinegar of roses, 1 ounce
juice of scented apples or lemon, ½ ounce
cloves, 1 drachm; *saffron,* 1 scruple.

Mix for a poultice for the pulse and the heart. Felix Platter, *Praxis medica,* p. 481.

Take of: Waters of balm distilled with wine,
borage, bugloss, of each 2 ounces
carbuncle stone, 1 ounce
species cordiales temperatae, 1½ drachms
powder of saffron, ½ scruple.

Mix. Make a epithem; place on the heart when the cause is cold.

Take of: Flowers of citron, fresh or dried,
red roses, borage,
bugloss, clove carnations,
marigold, of each 1½ pugils
saffron, 2 scruples
leaves of lemon balm, ½ handful
seeds of balm, basil, of each ½ scruple
species cordiales temperatae, 1 drachm
yellow part of fresh citron-rind, 3 drachms.

With a cloth, make quilted cordial sachets according to Art.[1] Besprinkle with spirit of roses or water of carbuncle stone.

1 Original wording: "*sacculi interpassati cordiales*". Blancard (1708) states that medicinal

1[2] Heart Palpitations

Travelling armies, because they go out of track and are arrogant, thus are generally unpleasant.

Similarly, moving air infiltrating into the residence of the heart from the womb, spleen or stomach, because it is irregular, is unpleasant and is expelled impetuously by the vigorous movement of the heart (which often touches the ribs).

And heart palpitation is the irregular and too impetuous movement of the heart, caused by the expulsive capacity irritated because of vapours, rather abundant liquid in the pericardium, worms, or the stone, a malignity, etc. What is heart palpitation?

The following water is greatly beneficial:

Take of: Pig's hearts recently removed, 3 in number
deer's hearts recently removed, 2 in number.

Cut into small pieces, add:
flowers of fresh borage, clove carnations,
balm, of each 1½ ounces
rosemary-flowers, fresh citron,
fresh orange,
Austrian saffron, of each ½ ounce
leaves of balm, 1½ handfuls
seeds of basil, avens,
carrot, citron, of each 3 drachms
yellow rind of orange, 1½ drachms
mace, 2 drachms
all the kinds of sanders, of each 2 scruples
waters of borage, bugloss, balm,
marigold, of each 1 pound
potent white wine, 1½ pounds.

Make an infusion according to Art for 24 hours, then distil well with a water-bath and add to the distilled product:
confection of kermes-berries with or without musk, 6 drachms.

Leave them in infusion in a closed glass until the red tincture is extracted, filter and store for use.

Let it be marked: **Sonderliches Wasser gegen das Herz-klopfen** ["Special water against heart palpitations"].

---

sachets, which contained powdered herbs etc., "are to be sewed or quilted down in several places, that the Ingredients run not together in a Lump."

2 There is no section 1 in the original book.

Or

Take of: Compound water of hart's-heart, 2 ounces
waters of tender hartshorn,
balm with wine,
cinnamon, borage,
carbuncle stone, orange flowers,
citron, scented roses, of each ½ ounce
confection of kermes-berries, 1 drachm.

Leave in infusion in a well-closed glass until the tincture is extracted. After decanting, add:

tincture of coral,
with prepared spirit of hart's-heart, 1½ drachms
quintessence of saffron, 15 grains.

Mix.

Let it be marked: **Herzwasser** ["Water for the heart"].

Against the palpitation and trembling of the heart, the famous Möbius in *Anatomia camphorae*, ch. 17, p. 79, gives some drops of oil of camphor in water of borage, or bugloss, and anoints the region of the heart with an ointment, pomade or with unsalted butter, or with oil of roses, each mixed with *saffron* and camphor.

If the cause is a malignity:

Take of: Mattioli's oil of scorpions, ½ ounce
juice of lemon, 2 drachms
best treacle, 1½ drachms
species cordiales temperatae, 15 grains
camphor dissolved with lemon juice,
extract of saffron, of each ½ scruple
oil of citron, 5 drops.
wax, a little.

Mix. Make an ointment.

In cases of the heart trembling:

Take of: Leopard's bane, bone of hart's-heart,
both [kinds of] behen, of each 1½ drachms
the two [kinds of] sanders,
spodium, wild vine,
both corals, of each 2 drachms
roses, 3 drachms
barley-meal, 3½ ounces
saffron, 1 scruple

oil of saffron, ½ ounce
bugloss-water, 1½ ounces
malmsey, 1 cupful
resin, 2 drachms.

Make a plaster. Ferdinando Epifanio, *Empyrica*, p. 417.[3]

3 *Empyrica* was written by Nicolaus Epiphanius in 1571, and not by Ferdinandus Epiphanius [Ferdinando Epifanio], who lived 1568–1638.

CHAPTER XV

Diseases of the Stomach

Often, the stomach is troubled by such a strong burning sensation that you would swear it is scattered with burning coal. If the stomach of a man longs for, and is filled with, the same things as the Eme bird from Japan, perhaps we should be less astonished: that bird is used to swallowing down not only each of the edible things given to it, but also fiery coals, thus Johann Pincier, book I, *Aenigmata* 30, p. 81.[1]

What is heartburn?

Heartburn is the rising up of sharp warm vapours, caused by defective fermentation in the stomach, or by bilious warm sharp humours.

Philip Grüling's round tablets with *saffron* described in book III, *Medicinae practicae*, ch. 7, fight strenuously against this condition.

Or
Take of: White chalk, nutmeg, of each 2 drachms
saffron, ½ scruple
sugar, as much as suffices, dissolved in mint-water.
Make round tablets. Müller, *Miracula et mysteria chymico-medica*, p. 83.

Pierre Potier, *Pharmacopoea spagyrica*, p. 552, presents another remedy as extremely effective in cases of heartburn and headache:
Take of: Chemically-prepared opium, ½ ounce
diaphoretic antimony, 2 drachms
ambergris,
eastern saffron, of each ½ drachm
musk, 1 scruple.
With syrup of poppy, form a mass. Dose: 7 to 10 grains.

1 The word "Eme" appears to have first come into the European vocabulary via the Dutch language in the last years of the sixteenth century. It was a borrowed term from Javanese, and means a cassowary (Lach, 1977, p. 549). An eme bird was given by the Javans to sailors of the first Dutch expedition there, with the claim that the bird could eat fire (McClymont, 1920, p. 21 et seq.). This cassowary was taken back to Amsterdam alive in 1597. An illustration of an eme bird occurs in Hulsius (1625, p. 43, entitled *Wunderbaren Vogels Eme*). The three species of cassowary (*Casuarius* sp.) are endemic to New Guinea and neighbouring areas, but not Japan.

 | DOI:10.1163/9789004435292_018

II[2] Anorexia, or Inappetence

The treachery of those who rise against the ruler responsible for their greatness is very disgraceful. It is a sign of an ungrateful heart and of dishonesty, a disgraceful offence, and a source of perfectly right indignation and upset on the ruler's side; and conspiracies have always caused misfortune for the rulers and those related to them.

Thus, sometimes the other body parts are displeased at the stomach because they procure all the things for the stomach with their care, toil and work, while it does not take part in anything else but enjoying the delights given to it. They plot together that the hands should not bring food to the mouth and that the mouth should not take the food given, and the teeth should not chew it. But while they wish in this way to tame the stomach, the conspiring limbs will soon come to extreme wasting. Daily experience confirms that, when the stomach lacks appetite, then the body gradually wastes away.

Inappetence is the loss of appetite because of weak fermentation in the stomach and is caused by redundant humours in the stomach, hindering the contraction of the membrane surrounding the stomach.

What is anorexia?

Our tincture for the stomach, whose recipe follows, is proven:

Proven tincture for the stomach

Take of: *Saffron*, myrrh, aloes, of each ½ drachm
rind of orange, 2 drachms
nutmeg, 1 drachm
root of cuckoo-pint, 2 scruples

conserve of clove carnations, 1½ drachms; the complete confection of kermes-berries, ½ ounce; Mattioli's elixir of life, 10 ounces. Moisten the dry ingredients with philosophical spirit of vitriol, 2 scruples, and leave them in infusion in a closed glass vessel, so that all the tincture is extracted. Then strain the tincture, and filter it, adding:
essence of liquidambar, ½ scruple.

Keep in a glass container.

Fabry von Hilden, *Chirurgicae observationes*, 32, *centuria* 5, p. 314, uses the following to cut into phlegm:

Take of: Cinnamon, *saffron*,
mace, of each 2 drachms
sugar, 4 ounces.

Mix. Dose: 1 spoonful with broth.

2 There is no section I in the original book.

Or

Take of: Cinnamon, liquorice, of each 1 drachm
galangal, ½ drachm
prepared coriander, 1 drachm
nutmeg, cubebs, of each 1 scruple
roots of peony, orris, of each 1 drachm
tops of hyssop, wild marjoram, 4 pugils
seeds of fennel, anise, of each 2 drachms
saffron, ½ scruple
ginger, 1 scruple
best sugar, 1 ounce.

Mix. Make an extremely fine powder.

Let it be marked: **Triseneth in anfang der Mahlzeit** ["Powder for the beginning of the meal"]. Rolfinck, *Methodus specialis*, book IX, section 1, ch. 2.

Additionally, to strengthen the stomach and the heart:

Take of: Oils of mint, scorpions, of each 3 ounces
nutmeg, 6 drachms
cloves, 2 scruples
finely powdered eastern saffron,
burnt and prepared hartshorn, prepared pearls,
confection of kermes-berries, of each 1 drachm.

Mix well in a mortar, make a liniment, store in a glass vessel, and smear very often in the region of the heart and stomach. Fabry von Hilden, *Observationes chirurgicae*, p. 314.

III Hiccups

The oracle and doctrine of Apollo, divine leader, is not truer than this, nor is the fact that Mercury has four faces firmer: a convulsion or hiccup supervening after a copious discharge of blood is bad, *Aphorism* 3, section 5, and *Aphorism* 4, section 5. Thus, also a convulsion or hiccup happening after excessive purgation is bad. Hiccup and red eyes after vomiting are also bad, *Aphorism* 3, section 7. And it is not good if it occurs when old persons have been purged excessively, *Aphorism* 41, section 7.

What are hiccups?

Hiccups are the distorted movement of the upper orifice of the stomach, caused by the excretive capacity being irritated because the longitudinal fibres of the stomach are weakened, by worms, cold pituitous humours, bilious humours, emptiness, raw fruit, air, a malignity, etc.

In cases of hiccups due to the stomach being full, from a cold cause, the emetic mentioned above, in the chapter on the diseases of the heart, § 2, is good for expelling the harmful humours from the stomach.

And if the problem persists, the humours should be prepared:

Take of: *Hydromel of saffron*, 2 ounces
oxymel simplex, 1 ounce
syrup of citron-rind, ½ ounce
Mix.

Bernard de Gordon's electuary with *saffron*, described in the chapter on hiccups, is also worthy of praise.

The following is good too, if the problem persists:

Take of: Mastic, fresh roses,
spikenard, *saffron*,
seeds of endive, of each ½ ounce.
Mix. Feynes, *Medicina practica*, book IV, ch. 8, p. 481.

IV Nausea

Wishing to eat chestnuts, as soon as they are found, is unseemly.[3] It is, indeed, extraordinary that our forefathers, when they inhabited the beautiful valley with that superabundant excess of things to eat, chose only apples, and they tasted them in an impious attempt against the command of GOD. For indeed those apples, created fit to bring ruin to soul and body, must have caused nausea.

Nausea is the vain urge to vomit the things contained in the stomach. Francis Bacon, Viscount of St Albans, *Historia vitae et mortis*, p. 121, applies a sachet of *saffron* to remove the sickness and nausea of sea-travellers. What is nausea?

3 Here we assume Hertodt means unripe chestnuts, eaten raw.

CHAPTER XVI

Diseases of the Intestines

I Hernia

Theatrum vitae humanae

At one time the weakness of the Sybarites was so great that in a town, they did not allow any work that was carried out with noise. Hence, it was not permitted to keep a cockerel, so that there was nothing that disrupted sleep. A Sybarite walking across the fields confirmed this weakness, as, when he saw some people ploughing, he said that he was getting a hernia or rupture by looking at them. A hernia is a protruding swelling of the intestines or of the bowels or of both, following the relaxation of the adjoining peritoneum. It is caused by chyliferous humours disseminated in the abdomen and in the whole body through the lymphatic ducts, and the very peritoneum which loosens and breaks, or by a shout, withheld breath, a fall, etc.

What is a hernia?

With this condition, when air flows in excess, the following clyster is good:

Take of: The ordinary carminative decoction, 6 ounces
electuary of bay-berries,
hiera picra, of each 6 drachms
extract of saffron, 10 grains.
Mix. Make a clyster.

Van Foreest prescribes these pills:

Take of: Cinnamon, *saffron*,
cubebs, mastic,
asarabacca, spike, of each 2 scruples
best aloes, 1½ ounces.
Pulverise the ingredients individually, and with syrup of roses, as much as suffices, make laxative pills. Dose: 1 drachm an hour before dinner.

II Worms

When a man dies, he shall inherit creeping things, beasts and worms, *Ecclesiastes*, ch. 10. But we experience the truth of this when we are still alive, as entire colonies of many worms, and the causes of almost unforeseen diseases, dwell in our body often also without us knowing.

 | DOI:10.1163/9789004435292_019

They are animals in the intestines or in another part of the body, originating from putridity because of corrupted humours. What are worms?

The following powder is a specific to kill and remove worms. Specific powder to kill worms

Take of: Root of jalap tinted with *saffron*,
dulcified mercury,
powder of vipers,
hartshorn alchemically prepared,
powder of coralline, of each 1 scruple
saffron, myrrh, of each ½ scruple
rectified oil of hazel-wood, 6 drops.

Mix. Make a powder. Dose: 1 scruple to ½ drachm. Take during the waning of the moon, when Friedrich Hoffmann, in the extremely erudite *Methodus medendi*, recommends a Thursday above the other days for the successful expurgation of worms.

Additionally, anoint the navel with this liniment:

Take of: Powder of aloes, myrrh,
saffron, of each 2 scruples
ox gallbladder, 4 scruples
powder of colocynth, ½ drachm
inspissated juice of wormwood, 1 scruple.

With oils of wormwood and Duchesne's diacolocynth and wax, as much as suffices, make a liniment.

Alternatively, apply the following to the navel:

Take of: Oil of rue,
aloes, coralline, of each 1 ounce
herb of wormwood, mastic, of each ½ ounce
prepared coriander, 3 drachms
finest wheat flour, 3 ounces
lupin-meal, 1 ounce
saffron, 1 drachm.

Make a powder with all these ingredients, and incorporate them with the abovementioned oil. Add: gall of grayling, as much as suffices.

Alternatively, first anoint the belly with butter and honey, then, above this layer, apply the following cerate:

Take of: Choice treacle, 1 ounce
bruised earthworms, 1 drachm
prepared coriander, 4 scruples

bull's gall, 2 scruples
myrrh, aloes,
saffron, of each 1 scruple
keenest vinegar, a little.

Mix. Incorporate over a slow fire and apply in the form of a cerate. Feynes, book IV, *Practica*, ch. 30.

Take of: Mithridate, 1 ounce
myrrh, aloes,
saffron, of each 1 drachm.

With wax, pitch and oil of bitter almonds, make a cerate. Ibid., p. 579.

Or

Take of: Roman coriander, 1 drachm
aloes, red coral, of each 2 drachms
wormwood, ½ ounce
mastic, 3 drachms
prepared coriander, 1 ounce.

Make a powder, incorporate with 3 ounces finest wheat flour; 6 drachms lupin-meal; *saffron*, 1 drachm; juice of rue, 2 ounces; oil of amber, 3 ounces. Make a plaster for worms. Place on the stomach. Epifanio, *Empyrica*, p. 411.

III Blind Haemorrhoids

To the present day, a question has been deliberated upon among doctors: whether aloes opens the orifices of the veins, and in particular those of the veins in the anus, or if, instead, it arrests their profusion.

Do aloes open the haemorrhoids?

The third author especially, and others, are of the first opinion: the ancient Serapion, *De simplici medicina*, ch. 201; Mesue, *De medicina benedicta*, ch. 1; Falloppio, *De simplicibus medicamentis purgantibus*, ch. 2; Fernel, *Methodus medendi*, ch. 7 and Della Scala, *Ad Cardanum excercitationes*, 160.

In the same way, Dioscorides, *De re medica*, ch. 2; Pliny, book XXVII, ch. 4; Galen, book VI, *De simplicium medicamentorum facultate*, ch. 25, and others, as well as Giovanni Manardo, Giovanni Battista Teodosi, Leonhart Fuchs, Francisco Vallès, Jacques Despars, Sebastien de Monteux, Jacques Dubois and others, think the opposite.[1]

1 In the original book, Manardo (Manardus) is misspelled as Monardus and Monteux (Montius) is misspelled as Montanus.

But this quarrel between learned men is easily resolved if we said that aloes externally applied to veins stops the blood flowing from them; when taken internally, instead, it opens the haemorrhoidal veins.

Blind haemorrhoidal veins are tumours and swellings of the veins of the anus, caused by stagnation because of the abundance and density of blood, and by the stretching out of the extremities of the veins. Karl Widemann prescribes this against the pain and the burning sensation caused by the haemorrhoids: What are blind haemorrhoids?

Take of: Oil of sweet almonds, oil of sesame,
fresh butter, oil of roses, mullein,
marrow of ox-leg,
oil of goose, hen, of each 1 ounce
egg yolks, 14 in number
eastern saffron, 4 scruples (which is very beneficial)
Thebaic opium, 10 grains
white wax, 2½ ounces.

Make an ointment, according to Art. Apply hot. *Tractatus de peste*, p. 131.

Or

Take of: Hens' grease, ½ ounce
pulp of apple cooked in ashes, 1 ounce
eastern saffron, ½ ounce
ointment of poplar-buds, ½ ounce
egg yolk, 1 in number.

Make a liniment. Rivière, *Praxis medica*, as well as Müller, *Mysteria et miracula chymico-medica*, p. 87.

Or

Take of: Seed-mucilage of psyllium
and quince, extracted with mullein-water, of each 6 drachms
powdered saffron, ½ drachm
egg yolk, 1 in number

oils of toad-flax, and beetle, and of wax, as much as suffices.

Make a liniment.

Or

Take of: Oil of roses, 1 ounce
egg yolk, 1 in number
saffron, 1 drachm
fresh butter, 1½ ounces
woman's milk, 2 ounces.

Make a liniment. Egg yolk and oil of roses exactly co-mixed, add *saffron*. When the intensity of the pain is high, add opium, 3 grains. Press a little in a leaden mortar. Make an ointment. Feynes, *Medicina practica*, p. 585.

When the pain is intense:
Take of: Ointment of poplar-buds, ½ ounce
oil of poppy, 1 ounce
opium, 1 scruple
saffron, ½ scruple.
Mix with whole egg.

Or
Take of: *Saffron*, myrrh,
lycium, of each 1 scruple
opium, ½ drachm; oil of roses,
mucilage of psyllium, egg yolk, as much as suffices.
Make a soft ointment. Epifanio, *Empyrica*, p. 481.

For more ointments with *saffron* to soothe the pain see Feynes, *Medicina practica*, p. 585.

Or
Take of: Sweet apples, cut into pieces
after removing the cores.
Boil in oil of linseed or violets, add an equal quantity of whitest breadcrumbs, bruise all these ingredients together and make them into a cataplasm. Add egg yolk and *saffron*, 1 scruple. Ibid., p. 586.

The following is excellent:
Take of: Chamomile, root of marshmallow,
melilot, mullein, of each 1 handful.
Boil until it thickens, add:
hardboiled egg yolks, 2 in number
saffron, myrrh, aloes, of each 1 drachm
meal of fenugreek, linseed, of each 6 drachms
unsalted butter, as much as suffices. Make a plaster.
One drachm of opium can also be added. Epifanio, *Empyrica*, p. 414.

IV Bleeding Haemorrhoids

Those who suffer from bleeding haemorrhoids rarely sleep and spermatize little. Albert the Great, *De animalibus*, book XV, ch. 7.

Bleeding haemorrhoids are the opening of the haemorrhoid veins caused by the abundance of blood or its pungency and sharpness.

What are bleeding haemorrhoids?

Besides venesection, which is superior to almost all the other remedies, Daniel Angelus recommends the following:

Take of: Seeds of quince, 2 drachms
psyllium, 1 drachm.

Shake with a sufficient quantity of plantain-water and make a mucilage, to which add:

extract of opium dissolved in a sufficient quantity
of goats' milk, 1 drachm
finely powdered saffron,
frankincense, myrrh, of each 2 scruples
colcothar,
dissoluble sugar of lead,
crocus martis, camphor dissolved in oil
of eggs, of each 1 drachm.

Mix. Smear on the affected part. *Appendix de compositione medicamentorum*, ch. 4, p. 37.

V Dysentery

It is beyond doubt that the miasmas of dysenteric excrements infect others.

Perhaps this was foreseen by the Hebrews, *Deuteronomy* ch. 27, "You shall have a place outside the camp, to which you shall go to relieve yourself, bringing a paddle on your belt, and when you relieve yourself, you shall dig a hole and cover up wholly that which has come from you with earth."[2]

Cats imitate this very thing by nature, and, with as much endeavour as possible, they either bury their excrements with gathered soil if there is abundance of it at hand, or they hide them with some other thing.

Dysentery is a frequent and bloody defecation, with pain in the bowels and in the intestines, accompanied by stool which is at times frothy, at times racemose. It is caused by sharp humours, a malignity, raw and spoiled produce, etc.

What is dysentery?

2 The correct reference is *Deuteronomy* ch. 23.

With this condition, nothing is found to be more efficacious than the very *essence of saffron*, see Zwelfer, *Pharmacopoeia regia*, p. 231:

Take of: Extract of tormentil,
caput mortuum of rhubarb, of each 1 drachm
sponge, pitch, moistened and burnt in a pipkin,
burnt cuttlefish-bone,
hardened eggshell, the inner membrane
removed, of each 1 scruple
extract of saffron, 1 scruple.

With tincture of bloodstone or emerald or Prévost's astringent liquor, make pills the size of small peas.

Or

Take of: Hartshorn chymically-prepared,
terra sigillata, of each 3 drachms
root of tormentil, bole armeniac,
jaw of pike [fish], of each 2 drachms,[3]
seeds of sorrel, plantain,
mace, of each 1 drachm
saffron, 1 scruple.

Mix. Make a powder. Thomas Bartholin, *Cista medica*.

VI Tenesmus

A disease is dangerous for us, because it has nothing familiar to us, but is entirely foreign. Plotinus, *Enneads*, 5, book VIII, ch. 11.

Heterogeneous humours are dangerous and foreign to nature, which always tries to expel them in every possible way as they are not internal and familiar.

What is tenesmus?

This happens with tenesmus, which is the constant urge to have a bowel movement with pain, as nothing or a small amount of slimy or somewhat bloody stool is evacuated. It is caused mainly by salty humours or by excessive refrigeration or by intestinal worms.

3 Writing slightly earlier in the seventeenth century, Izaak Walton, 1653, p. 146, confirms the medicinal value of the pike, writing, "the bones, and hearts, & gals of Pikes are very medicinable for several Diseases, as to stop bloud [blood], to abate Fevers, to cure Agues, to oppose or expel the infection of the Plague, and to be many ways medicinal and useful to the good of mankind …".

In cases of extremely persistent tenesmus, Brunner in *Consilia*, 50, p. 287, applies this suppository:

Take of: Lycium, frankincense, *saffron*,
gum Arabic, of each ½ drachm
opium, 2 scruples.

Make suppositories with a string. Do not leave more than one hour in the intestine.

Or

Take of: Lycium, *saffron*, of each 1 drachm
ceruse washed in plantain-water, 2 drachms
fat of capons' kidneys, as much as suffices.

Make a suppository and attach a string to it, so that when an hour has elapsed, it can be taken out again.

Or

Take of: Frankincense, myrrh, lycium, of each 2 drachms
opium, 1 drachm
saffron, 1 scruple.

With mucilage of psyllium and egg yolks, as much as suffices, make small and fine suppositories. Ferdinando Epifanio, *Empyrica*, p. 476.[4]

VII Diarrhoea

It is incredible what is said about the raven: that it was condemned to thirst, for as long as the fig is milky, because of a lie it said. And, hence, its frequent croaking is a sign that it is going to rain. It is beyond question that ravens are thirsty, and not without cause, as for the sixty days of summer, during which they are afflicted with diarrhoea, they stay away from drinking, in order to prevent that inconvenience, and they stop the bowels.[5]

Abstinence from drinking is, in fact, a remarkable remedy for diarrhoea, which is the frequent flux of the bowels with or without pain, caused

What is diarrhoea?

4 Again, the author of *Empyrica* was Nicola Epifanio, not Ferdinando.

5 Aelian, I. 46, wrote, "All through the summer the Raven is afflicted with a parching thirst …" this being a punishment by Apollo for disobedience, and "Ravens know that in summer they suffer from looseness of the bowels; for that reason they are careful to abstain from moist food" (book II, 51). In German folklore, ravens are condemned to thirst in August, because they alone, of all the birds, did not mourn the death of Christ (Köhler, 1898, p. 3).

by worms, or cold air, spoiled milk, spoiled food, catarrh, or sharp or salty bilious humours.

In cases of symptomatic diarrhoea, premised the appropriate evacuation of the harmful humours, Stocker in *Empyrica*, p. 100, inserts this suppository which makes the affected person costive by cleansing the body.

Take of: *Saffron*, myrrh, frankincense, of each 2 scruples
lycium, hypocistis, acacia, of each 2½ scruples
choice mastic, 1 scruple
opium, ½ scruple.

With a sufficient quantity of seed-mucilage of psyllium, make small suppositories the length of a digit, with a string attached. Successively insert into the anus.

VIII Colic

An old precept: if someone, while travelling, has the winds against him, his affairs advance with difficulty. Not seldomly it is observed that this is especially true in war.

Our life is a journey, and no-one is born with Jove so propitious that he does not have to struggle with the many enemies of our body, often, alas, hostile signs. Among these, the colic is not the least, as it torments day and night and it tortures and consumes until the end, generally accompanied by hostile winds.

What is colic?

Colic is, in fact, a pain in the intestines from an unpleasant movement of the fibrils of the large intestine, reaching the brain and the sensorium. It is caused by mucous tartar impregnated with salino-nitrous particles, by flatulency, worms, phlegm, stones in the bladder, tumours, or astringent raw food, etc.

Besides our very simple *clyster of saffron*, mentioned above in the chapter on the uses of preparations of *saffron*, this too is remarkable and anodyne:

Take of: Root of marshmallow, white lilies, of each 1 ounce
flowers of Roman chamomile,
mullein, mallow,
elder, of each 1 handful.

Boil in a sufficient quantity of milk, in 8 ounces of colature dissolve:
extract of castoreum,
saffron, of each ½ scruple
egg yolk, 1 in number.
Mix. Make a clyster.

In place of the these, internally, a purifying *saffron*-flavoured broth of wine can be taken. See its recipe in my *Tartaro-Mastix Moraviae*, p. 60.

Otherwise, *spirit of saffron*, both simple and compound, used internally and externally, is of remarkable service with this condition, as well as with other conditions.

The *compound spirit of saffron* is this: *Compound spirit of saffron*

Take of: *Common extract of saffron*, 1 ounce
tincture of tartar,
salted spirit of sal-ammoniac, of each 2 ounces
gold leaves, 60 in number
confection of kermes-berries, as much as suffices.

Mix.

Feynes, *Practica*, book IV, ch. 22, describes extremely powerful *pills of saffron*.

Or

Take of: Intestines of wolf washed in potent wine
and dried in a small, portable oven not fully burning, 2 drachms
burnt hartshorn washed in wormwood-water, 1 drachm
cumin, galangal, of each 2 scruples
dried wolf dung, cinnamon, of each ½ drachm
pepper, ginger, *saffron*, of each ½ scruple
sugar, half the weight. Ibid., p. 533.

Alternatively, the following powder, which is a specific:

Take of: Testicles of horses, washed in wine Specific powder
and dried out in a small, portable oven,
intestines of wolves prepared in the same way,
dung of the same which has not touched soil, of each 1 scruple
powder of saffron, ½ scruple.

Mix. Make a powder for 2 doses. Take in a convenient vehicle.

Or

Take of: Rind of sweet orange,
seed of anise,
our round tablets of saffron, of each 1 drachm.

Mix.

Let it be marked: Pulver gegen das Reißen ["Powder against the colic"].

Or

Take of: Confections of anise, fennel,
caraway, cumin,
confectio sine anima of saffron, of each 1 drachm.

Mix in a box.

Let it be marked: Confection gegen das Reißen ["against the colic"].

Take of: Syrup of mint,
waters of mint, fennel, of each 1½ ounces
Mattioli's quintessence of the Augustan dispensatory, 6 drachms
spirit of saffron, ½ drachm
saltpetre, 10 drops [more likely to be 10 grains]
Duchesne's nepenthes, 2 grains. Mix.

Let it be marked: Mixtur gegen das Reißen / löffelweis zunehmen ["Mixture against the colic/to be taken in spoonfuls"].

Water specific for the colic

We make public a water specific for the colic, so far kept for ourselves, whose recipe is this:

Take of: Cleaned dung of doves,
seeds of anise, of each 4 handfuls
best wine, 4 measures.

Digest for several days, add:
yellow rind of fresh sweet orange, 2 ounces
Austrian saffron, ½ ounce

Distil in a water bath, with *Austrian saffron*, ½ scruple, and if you want, musk, 5 grains, suspended from the beak of the alembic. Keep in a tightly closed glass.

Let it be marked: Anti-colic water.

Or

Take of: The abovementioned anti-colic water, 1 ounce
spirit of zedoary, 7 drops
volatile salt of amber, 5 grains.

Mix.

Let it be marked: Wässerle gegen das Reißen ["Water against the colic"].

Additionally, when extremely intense pain troubles the patient:

Take of: Leaves of pellitory of the wall, henbane, of each 1 handful.

Boil and bruise, and to the bruised ingredients add:
powder of flowers of chamomile,
seed-meals of linseed, fenugreek, of each ½ drachm

saffron, ½ drachm
opium, ½ or 1 scruple
oil of poppy, 2 ounces
chamomile, dill, of each 1 ounce.

Make a cataplasm, but apply with caution. Sennert.

Or

Take of: Gruel of rye-meal prepared with milk, 1 pound
add:

powder of saffron, 2 drachms
oils of rue, 1 ounce
chamomile, dill,
pimpernel, the 7 flowers, of each ½ ounce.

Mix.

Our very own gruel for babies

Let it be marked: **Köchel gegen den reißen über den Leib warm auffzulegen** ["Gruel against the colic, to be applied warm over the body"]. Our very own gruel for babies.

Take of: Root of henbane (which is also said to help
if hung around the neck), 1 ounce
white poppy heads, 4 in number
mandrake-apple, or root of the same, ½ ounce
dill, flowers of chamomile, of each 2 pugils
spike, 1 drachm
saffron, ½ drachm.

Make a decoction in water, boiled until a quarter of the water has evaporated. Bathe the affected part. Feynes, p. 536.

CHAPTER XVII

Diseases of the Liver

1 Inflammation of the Liver

It is more profitable to end the war and pack up, while there is still some hope for us to carry on and get away. Thucydides, book VII, on Nicias' obstinacy in Sicily.[1]

Inflammation of the liver must be stopped immediately, and one must take care that it does not go into suppuration.

Galen, *Methodus medendi*, ch. 5

The inflammation of the liver is indeed a dangerous and extremely deadly illness.

What is inflammation of the liver?

Inflammation of the liver is a hot swelling of the same, caused by blood stagnating in the liver tissue, with constant fever, pressing pain and a sensation of weight in the right hypochondrium.

To suppress it:

Take of: Leaves of wormwood, celtic nard,
asarabacca, *saffron*, of each 2 drachms
orris, myrrh, storax,
bdellium, gum ammoniac, of each ½ ounce
marrow of calf-leg, 1 ounce
cerate of mastic-tree, 1 pound.

Mix. Feynes, *Practica medica*, p. 459.

To relax, soften, and strengthen:

Take of: Aloes, storax, of each 1 ounce
mastic, ½ ounce
flowers of chamomile,
melilot, of each 6 drachms
madder, asarabacca, sweet flag, of each ½ ounce
seeds of smallage, parsley, of each 2 drachms
spike, *saffron*, of each 1½ drachms
roses, 2 scruples

wax and gum ammoniac, of each 3 ounces; propolis, 1½ ounces; oils of lilies, orris, of each, as much as suffices; wax liquefied in oil; then add propolis

1 Nicias, c. 470–413 BC, was an Athenian politician.

 | DOI:10.1163/9789004435292_020

and gum ammoniac dissolved in vinegar to these. Mix the remaining ground ingredients and make a plaster according to Art. Apply to the part; it dissolves the swelling. Feynes, book IV, ch. 12.

If it tends to suppuration:
Take of: Decoction of roots of marshmallow, ½ pound
barley-meal, 2 ounces
seeds of linseed, fenugreek, of each 1 ounce
fats of geese, ducks, of each 1½ ounces
English galingale, schoenanth, of each ½ ounce
spike, *saffron*, of each 1 drachm
gum ammoniac dissolved in oil of lilies and wax, of each 1 ounce. Prepare a plaster: melt the fat and wax in oil, add the bruised marshmallow, then the gum ammoniac, and finally the powders. Ibid., p. 460.

II Hepatitis, or Imbecility of the Liver

Neglect to investigate suspects, and excessive confidence, destroyed many rulers. Many patients suffering from imbecility of the liver die, dropsy having followed, because the doctor has not been summoned quickly, since, having too much confidence in themselves, the patients are solicitous about removing the external cause, as well as the internal, with their own strength.

Pierre Grégoire, *De republica*, book XXII, ch. 11

Imbecility of the liver is the damaging of the faculties of the same. It is caused by a cold distemper, or by excessive blood evacuation, refrigerants applied to the liver, or excessively heating, which dissipates the natural heat.

What is hepatitis?

The following powder is extraordinarily beneficial:
Take of: Intestines of hens,
with the internal layer of the stomach separated from the flesh and washed in white wine, 2 ounces, burn in an oven and quench with white wine and wormwood-water. Add to each ounce of ash: galangal, cinnamon, of each 1 drachm; *saffron*, 2 scruples; cloves, ½ drachm; prepared and powdered liver of wolf, 6 drachms; lozenges of sugar of roses, 1 ounce. Make a strengthening powder.

And the following electuary is no less efficacious:
Take of: Conserve of agrimony,
rob of raisins, of each 3 drachms
prepared liver of wolf, 1½ drachms

species diatrion-santalon
saffron, of each 8 grains.

With syrup of coral, make an electuary. Dose: 3 drachms, drinking after a decoction of agrimony.

When it is leaning towards cachexia:

Take of: Duchesne's powder for cachexia, 2 scruples or 1 drachm
our syrup of saffron, ½ ounce.

Take five hours before luncheon, drinking after hen's broth, in which some wood of sassafras and China with the aperient roots have boiled. At the end of the boiling add leaves of dodder; agrimony; common chicory; Roman wormwood; and a little *saffron*; some salt of wormwood; and lovage; and spirit of salts, a few drops. Take this, staying in bed for one hour, then getting up and walking for another hour, continuing for 15 or 20 days, with a cleansing of the spleen on the fifth day.

Or

Take of: Rob of elder-berries, 3 ounces
saffron, 3 drachms.

Mix. For a sudorific.

Additionally, in case of cold distemper of the liver:

Take of: Myrrh, mastic, of each 3 drachms
spike, English galingale, schoenanth,
dried eupatorium, spleenwort,
sweet flag, of each 2 drachms
saffron, gallia moschata, of each 1 drachm.

Knead the myrrh with the mastic in sweet wine and, adding the other ingredients, make a plaster and place on the liver. Feynes, *Practica*, book IV, ch. XI, p. 451.

III Dropsy

No fate is satisfied to hurt only once, since every day we see misfortunes come after misfortunes so that one adversity, spreading everywhere with its many roots, produces a forest of misfortunes. Similarly, also, is a poisonous tree: if even a little space is given to it in a garden or in a field, gradually spreading itself, it destroys all the nearby plants with its miserable and harmful propagation.

The lymphatic fluid in our body imitates the latter. If the more flexible particles of the same are shattered by the remaining more rigid particles, they cause an unnatural saltiness, from which soon other evils arise. This fluid, harmful not only because of the acquired irregular quality, but also because of its quantity, often breaks its ducts and, forsaking its abode, goes into the abdomen. It submerges all the viscera of the lower belly with its abundance and causes dropsy, so that, not improperly, with Plautus, we can proclaim it a sea of evils.

Dropsy is an unnatural swelling, caused by rather serous lymph driven away to the external parts and the extremities of the body or, if the lymphatic vessels have broken, collected only in the abdomen, or, finally, by air because of damaged sanguification. What is dropsy?

The *syrup of saffron-flowers* described above, in the chapter on the uses of preparations of *saffron*, is a specific and is extremely remarkable with every kind of dropsy, moist as well as dry.

Or

Take of: *Our syrup of saffron*, 6 drachms
extract of mechoacan, 1 scruple
magistery of jalap tinted with *essence of saffron*, 6 grains
salt of danewort, 7 grains
best cinnamon-water, 2 drachms
spirit of orange-rind, ½ scruple

Mix, make a draught, administer in cases of anasarca.

The following powder for anasarca, as well as for tympanites, is very much tested by experience: Tested powder for tympanites

Take of: Coagulated spirit of salt, 2 drachms.

Besprinkle twice with *spirit of saffron* and a third time with essence of the same and make a powder. Divide into 10 separate powders of equal dosage. Administer in *our anti-colic water of saffron*. Also efficacious against tympanites is *compound spirit of saffron*, up to 20 drops, taken in anti-colic water. Additionally, the swollen parts can be anointed with it, warm, after rubbing.

IV Jaundice (*Icterus flavus, aurigo, galgulus, arquatus, morbus Regius*)

A princely skill is saying good things, and devising evil with the thoughts.

Jaundice often imitates this, since if it arises through fevers on day 7, 9, 11 or 14, it seems to announce the benignity of the crisis; nonetheless, it is bad, since it is accompanied by the induration of the right hypochondrium. Hippocrates, *Aphorism* 42, section 6, "It is bad for a person suffering from jaundice that the liver becomes indurated."

What is jaundice?

Jaundice is the yellow pigmentation of the whole body, and of the tunica adnata because of the profusion of bile to every bit of skin, caused by the obstruction of the bile duct or of the cystic duct, or by the compression of the same.

Celandine and *saffron* have its signature. Croll in *Basilica chymica* and Johann Mylius, *Antidotarium medico-chymicum,* p. 70, state that they both benefit for their resemblance to the yellow pigmentation.

To the jaundiced, Philip Grüling prescribes the following claret with great benefit for the health:

Take of: Horehound, pennyroyal,
maidenhair, vervain, of each 1 handful
sweet flag, 2 drachms
eastern saffron, 1 scruple; choice cinnamon, 1 drachm.

With sugar and wine, make a claret.

Or

Take of: Greater celandine with everything, 1 handful
leaves and flowers of St John's wort, of each ½ handful
shavings of ivory,
powder of goose dung collected in the spring, of each 2 drachms
saffron, ½ drachm.

Insert the goose dung and *saffron* in a nodulus, and boil everything in equal parts of white wine and wormwood-water down to 1 pound. Make an apozem in three morning doses, which afterwards can be repeated several times. Rivière, and Hendrik de Roy.

Or

Take of: Roots of nettle, 1 pound
saffron ½ drachm

white wine, 2 pounds. Bruise and express the juice. For a dose: 4 ounces. Horst, *De tuenda sanitate,* p. 241.

Some praise the following powder as a remarkable remedy in case of jaundice, and they divide it into three parts and administer it for three days in the morning in a decoction of chickpeas and peas.

Take of: Seeds of columbine, shavings of hartshorn,
eastern saffron, of each 1 drachm

Make a fine powder. Divide into three doses. Platter, *Praxis medica*, p. 140.

Philipp Grüling speaks highly of this:

Take of: Powder of seeds of cress, 2 scruples
columbine,
eastern saffron, of each ½ scruple.

Mix with wine. Give in the morning for up to three days. *Praxis medica*, book III, part 5, p. 306.

Or

Take of: Powder of earthworms, 6 drachms
roots of strawberry, seeds of columbine, of each, ½ ounce
eastern saffron, 2 drachms
salts of chicory, wormwood, of each ½ drachm.

Make a powder. Dose: 1 drachm with wine of wormwood. Balthasar Timaeus von Güldenklee, *Descriptiones medicamentorum epistolae*, p. 459.

Or

Take of: Ox gallstone,
lesser centaury, nutmeg,
saffron, of each 1 drachm

Make a powder. Dose: 1 scruple. Administer in wine three times. Sennert.

The excellent Dr Andreas Graba, our Cephalus, has recently administered the following powder to an honest and pious wife:

Take of: Hartshorn philosophically prepared, 2 scruples
yellow stone found in gallbladders of bulls,
seeds of columbine, of each ½ drachm
salts of ash-tree, *saffron*, of each ½ scruple

Make a powder for two doses. Administer a portion mixed in wine of wormwood, water of lesser centaury, dodder, strawberry or decoction of hartshorn with roots of asparagus, syrup of eupatorium, chicory, and juice of citron. See *Elaphographia*, p. 147.

Powder against jaundice

The following powder has often been tested by experience:

Take of: Goslings' dung collected in spring,
powder of washed and dried earthworms, of each 2 drachms
coagulated spirit of salt,
prepared ceraunia stone, of each 1 drachm
fixed diaphoretic antimony,
eastern saffron, of each ½ drachms.

Mix. Dose: ½ drachm morning and evening. Administer in wine or decoction of horehound, or strawberry for a further internal cleansing and unblocking.

Saffron can also be added to cleansing medicaments designed for this condition, for instance:

Take of: Extract of greater celandine,
lesser centaury,
Austrian saffron,
Syrian powder, of each 5 grains
mass of Schröder's pills of tartar, 1 scruple.

With sufficient quantity of water of celandine, make nine pills.

Additionally, apply the following to the navel. It was tried by the wife of Achilles Gasser:

Take of: Honey, 2 drachms; *saffron*, ½ drachm
wheat-meal, 1 drachm.

Shake well together, and apply twice every day, warm, to the navel. Hieronymus Welsch, *Sylloge curationum et observationum medicinalium*, p. 14.

Observation on *saffron*

Some women put this mixture in the shell of a walnut and apply it to the navel of women in childbed, and they take it away within the same hour during which it has been applied, when they see that the shell is empty. They repeat the procedure two or three times, and generally do this for 14 days after the delivery, most certainly believing that all women in childbed must be afflicted with jaundice.

But who does not see that such jaundice comes from the dyeing of *saffron*? This paste, insinuating itself through the navel, mingles with the blood contained in the closest veins and, thus, according to the laws regulating circulation, through the veins is carried to the heart, and then through the arteries to the skin of the whole body.

To preserve from and cure jaundice in childbirth, some used to hang a nodulus, containing fresh *saffron*, camphor, and asafoetida, around the neck with a golden string. From a report via letter of our colleague, Dr Georg Sebastian Jung, PhD and MD.

V Obstruction of the Liver

When public danger impends, the opinions of those disagreeing with each other must be reconciled. Dionysius of Halicarnassus, book VIII.

Obstruction of the liver is a sign of considerable danger. In fact, it happens because bile does not mingle together with the pancreatic juice in the intestines and, with it retained, it is unavoidable that severe illnesses arise, if it is not cured promptly by reconciling them.

Obstruction of the liver is, in fact, the narrowing of the vessels in the liver, caused by phlegm, clotted and tartareous blood, a stone, or air filling the cavities of the vessels. What is obstruction of the liver?

When one cleansing is of no benefit, the following pills set every point free.

Take of: Hiera picra, 2 drachms
agaric, turpeth, of each 3 drachms
ginger, spike, zedoary, asarabacca, anise, cinnamon, *saffron*, juice of euphorbium, wormwood, of each ½ drachm; hepatic aloes, 3 drachms. Make pills with juice of smallage or oxymel. Feynes, *Practica medica*, p. 467.

Or

Take of: Mass of Duchesne's pills of gum ammoniac
(they contain *saffron*),
Schröder's pills of tartar,
extract of rhubarb, of each ½ scruple
resin of jalap tinted with *saffron*,
sal martis subdulcis, of each 5 grains
extract of saffron, 2 grains.
With juice of wormwood, make 9 pills wrapped in gold, for a dose.

Take of: Roots of bitter-sweet, 3 ounces
tops of hops, eupatorium, of each 2 handfuls
dodder, hepatica,
chicory with everything, of each 1½ handfuls.
Pour into whey of goats' milk, as much as suffices, to the height of 4 digits, for 3 days, then distil a water.
Take of: This [water], 3 ounces
syrup of the 5 roots,
of chicory with rhubarb, of each ½ ounce
Make an aperient decoction. In this pour:

white mechoacan, 1 drachm
rhubarb, ½ drachm; spike, 4 grains
saffron, 8 grains.
It will cleanse at once with moderation.

This powder is efficacious to expel the remains of humour through the urinary tract:
Take of: Madder, ½ ounce
celtic nard, sweet flag, of each 3 drachms
myrrh, asarabacca, valerian, rhubarb, seeds of common smallage, of each 2 drachms; dried liquorice, schoenanth, spleenwort, dried wormwood, dodder, greater germander or felty germander, of each 1½ drachms; *saffron*, 1 drachm.

Make into an extremely fine powder; this way the power will penetrate more quickly. Give 2 scruples with wine or administer with honey of an amount equal to the size of a nut. Feynes, p. 469.

Among topical medications, the plaster of sanders, once called emplastrum incognitum ["unknown plaster"], which *saffron* deservedly adorns, is suitable to restore an obstructed and indurated liver, to soften and to loosen the infarcts in it.

Avicenna praises another plaster for the hardness of the liver as wonderful:
Take of: Mastic, *saffron*,
gum ammoniac, of each 1½ ounces
oils of mastic, roses, of each 2 ounces
wine, 20 ounces.
Melt wax and oil, and mix the rest with it.

VI Distemper of the Liver

Julius Caesar, being asked how he had achieved such great things, replied, "not deferring anything."

A doctor could easily give himself a victory, if he does not give a break to the disease and attacks it quickly, before the distemper becomes habituate, especially if it concerns one of the main organs, one of which is in fact the liver.

What is the distemper of the liver?

The distemper of the liver is the recession of the same from its natural state and can be hot or cold.

When it is hot:

Take of: Powdered red sanders, 1 drachm
flowers of red roses, 1 handful; of white poppy, 2 drachms; species aromatica rosata, 1 drachm; *eastern saffron*, 1 scruple; camphor, ½ scruple; rose-water, wine vinegar, of each 3 ounces. Make an epithem. Job Kornthauer, *Tractatus de peste*, p. 18.

CHAPTER XVIII

Diseases of the Spleen

1 Obstruction of the Spleen

Emperor Trajan compared the spleen to the imperial treasury: in fact, if it swells because of an obstruction, all the other organs gradually waste away. And Alciato expressed this very beautifully in an emblem:

Humani quod est splen corporis, in populi re
Hoc Caesar Fiscum dixcrat esse suum :
Splene aucto reliquis sabescunt corporis artus :
Fisco aucto argitur civica pauperies.

["What the spleen is to the human body, the emperor said
His treasury was to the public wealth:
If the spleen swells, the other parts of the body waste away;
If the treasury swells, it is a sign of public poverty."]

What is obstruction of the spleen?

Obstruction of the spleen is the tightening of the vessels in the spleen, caused by thick tartareous humours, etc.

A universal remedy in cases of obstruction of the spleen is the *asthmatic spirit of gum ammoniac and saffron*, described in [ch. XIII] § 2, on asthma, taken in aperient waters or in an appropriate vehicle, to which some spirit of scurvy-grass can be added, if you like. And so is *our spirit of saffron*, mentioned in the chapter on the uses of preparations of *saffron*.

For cleansing:

Take of: Duchesne's pills of gum ammoniac, 1 scruple
Schröder's pills of tartar, 7 grains
resin of jalap tinted with *essence of saffron*, 5 grains.
With our *syrup of saffron*, make 15 pills wrapped in gold for a dose.

Extremely useful for the strengthening of the spleen is Solenander's electuary, described by Hartmann, to which *essence of saffron* is added with great benefit. Dose: the size of a chestnut.

Liniment of proven efficacy

Additionally, the following liniment is of proven efficacy in cases of obstruction of the spleen:

 | DOI:10.1163/9789004435292_021

Take of: Oils of capers, 3 drachms
of seeds of rocket by expression,
Roman coriander by expression, of each 1 drachm
oil of saffron by insolation, 2 drachms
rectified tartar, 15 drops
gum ammoniac dissolved in vinegar of squill, 2 drachms
inspissated juice of spleenwort, 1 scruple
extract of saffron, ½ scruple
ointment of marshmallow, ½ ounce
wax, as much as suffices.
Make a liniment.

11 Hypochondria

Great is the temper of a mind not being affected by anything pleasant or sad.

Socrates could pride himself of this: in all his life he never laughed or cried, and his countenance was one and the same to such a degree that it could be changed by neither nuptial joy nor the sadness of the day when he drank poison.

The immovable Socrates

But as much as a Socrates is uncommon, thus, even more uncommon are those who imitate him. In fact, those plagued by hypochondria in particular are so full of panic that the idle noise of a sling scares them like birds, and often they are troubled more by the thought than by the actual thing.

Hypochondria is the collection of dregs of argillous and mucilaginous tartar, impregnated with salino-nitrous particles in the branches of the mesenteric vein, but especially in those of the celiac artery and of the pancreatic duct, when acid fluid assembles, producing various symptoms.

What is hypochondria?

In his *Corallorum historia*, p. 166, Gans describes the anti-hypochondriac tragea of Johann Popp, which is as follows:

Take of: Salt of coral, pearls, of each ½ ounce
true unicorn's horn, 1 drachm; diaphoretic gold, 2 drachms
saffron, 1 drachm
oils of cinnamon, of mace, of each 12 drops.
Mix. Make a tragea.

Balthasar Timaeus von Güldenklee adds *saffron* to his essence and elixir for hypochondria as well as for the spleen. See *Descriptiones medicamentorum* in his *Casus* and *Epistulae*.

Finck in *Encheiridion dogmatico-hermeticum*, p. 24, describes the curative water of a famous doctor, extremely serviceable for hypochondriacal melancholy, which contains *saffron*.

Saffron is also one of the ingredients of King Sabor's syrup, which we use against all the melancholic conditions, as well as for strengthening the heart.

For the correct way to prepare this syrup, see Luigi Settala, book IX, p. 177.

Observation on *saffron*

Borel, *centuria* 2, observation 99, reports that a certain noble woman, who was afflicted with such a severe melancholy and oppressive sadness that she was taken by a constant urge of crying, was cured by carrying *saffron* on the heart.

Quintessence of saffron and spirit of the same, administered in water of borage, cure the symptoms of palpitation and tightness of heart deriving from hypochondria.

III Scurvy

Kircher

Everyone sees the various transformations that happen daily in the world. In fact, we know that a basilisk comes from a snake egg swallowed by a cockerel; that, in caves where eagles and vultures abound, dragons come from the putrefied seeds of various animals, of snakes, etc., a part of which remains in the corpses; that bees come from ox dung; a bruchus from deer's excrements. We know that savin comes from a little plant of cypress transported to the northern regions, and that hemlock comes from corrupted parsley.

It is well known that, with scurvy, many and diverse types of diseases and symptoms derive from blood deprived of volatile salt.

Various pains of the legs and calves, lumbago, insensibility, paralysis, tremor, convulsion, delirium, wakefulness, appetite loss, constant spitting, dropsy, jaundice, diarrhoea, constipation, difficulty in breathing, and an almost suffocating tightness of the chest all result from scurvy. Not to mention the spots on the feet, bleeding gums, gum and teeth infections, fevers, scorbutic gangrene, atrophy, and others.

What is scurvy?

Scurvy is, in fact, a cachexia caused by a particularly corrupted melancholic humour, afflicting with infirmity of the legs, spots, swelling of the gums, a bloodshot condition, looseness of teeth, and other symptoms.

With this condition, the following water is good:

Take of: Goat's milk whey,
Rhenish wine, of each 1 pound
green pimpernell, watercress, of each 3 handfuls

boil and strain, to 15 ounces, then add:
juice of scurvy-grass, 2 ounces
saffron, ½ drachm.
Mix.

When the patient is troubled by dyspnoea, the illustrious Dr Johann Michael makes use of the following:
Take of: Water of turnip-flowers, 3 or 4 ounces
Dornkreill's anti-scorbutic water, 2 ounces
syrup of conserve of elecampane root, ½ ounce
essence of saffron,
of elecampane, of each 1 drachm
the compound spirit of gum ammoniac, ½ drachm.
Mix.

Let it be marked: **Lufft-Safft auff etlich mahl**, ["Breath-syrup, for several times"] 2 or 3 spoonfuls.

Or
Take of: Essences of elecampane, *saffron*, of each 3 drachms
the *asthmatic spirit of gum ammoniac and saffron*, 2 drachms
the anti-scorbutic spirit, 1½ drachms.
Mix.

Or
Take of: Faeculae of cuckoo-pint, 1½ drachms
species diaireos, ½ drachm
magistery of scurvy-grass, 2 scruples
saffron, 1 scruple.
Mix.

Let it be marked: **Lufft-Pulver** ["Breath-powder"]. Dose: 2 knife points.

With this condition, salted spirit of sal-ammoniac, taken with *tincture of saffron* and water of scurvy-grass, is also good. Frederick Dekkers in *Notis ad praxim Barbettianam*, ch. 111, *De scorbutus*, p. 151.

The following julep helps in cases of scorbutic gripes:
Take of: *Our anti-colic water of saffron*,
described in the section about colic,
Dornkreill's anti-scorbutic water, of each 2 ounces
van Foreest's scelotyrbic syrup,

of rind of oranges, of each 1 ounce
spirit of saffron, 1 drachm.

Mix.

Or

Take of: Spirit of scurvy-grass, [water]cress, of each 2 drachms
quintessence of saffron, 1 drachm.

Mix.

Let it be marked: **Essenß gegen den Schmerßen / 30. 40. tropffen auff einmahl** ["Essence against the pain/30–40 drops at a time"].

The following tincture is greatly recommended against wandering pain of the joints in the back of the neck, in the legs and feet:

Take of: Spirit of earthworms, 5 ounces
sal-ammoniac, ½ ounce
scurvy-grass, 2 drachms

in which dissolve:
Venice soap, 2 ounces
camphor, 6 drachms
opium, 1 scruple
saffron, 1 scruple.

Mix. Digest for several days and nights, then filter, and store for use in a tightly closed glass. It can be applied with a linen cloth previously warmed. Frederick Dekkers in *Notis ad praxim Barbettianam*, pp. 150 and 151.

Or

Take of: The Augustan dispensatory's gout water, 2 ounces
the *compound spirit of saffron*,
worms, of each 6 drachms
scurvy-grass, cress, of each 3 drachms.

Mix.

CHAPTER XIX

Diseases of the Kidney and Bladder

1 The Stone

Hidden enemies are the worst, and concealed enmities are always to be feared more than those which are openly declared. Cicero, *In Verrem*

The stone often builds a hidden enmity, since, lurking in the kidney parenchyma, suddenly moving out of its orbit, it occupies the renal pelvis or obstructs the ureters. Then, after troubling with the most severe pain, it causes inflammations, epilepsy and even death.

What is the stone?

The stone is a solid substance of the same nature as a stone or tuff, originating by means of heat from mucilaginous tartar impregnated with salinonitrous particles, and with a *λιτοποιετική* [litopoietikí, "lithopoietic"] faculty, as others claim.

This powder is efficacious:

Take of: Crystals of saltpetre, 3 drachms
vitriolated tartar,
crab's-eyes, perch stone, of each 1 drachm
saffron, 2 scruples
mace, 1 scruple.

Mix. Dose: 1 drachm.

Or

Take of: Tartaric acid
calcined crystals, 1 drachm.

Bedew well with spirit of vitriol, as much as suffices, knead with a stone, bruise, and, when they are dry, add:
clusters of dog-rose hips,
ripe seed of fir, of each 1½ drachms
saffron, 2 scruples.

Make everything into a powder, store for use. Dose: for healing, ½ drachm to 2 scruples. Take in water of couch grass, for preservation, once or twice a week with light hen broth.

Or

Take of: Timaeus' lithontriptic magistery
from woodlice and prepared crab's-eyes, 1 scruple

© KONINKLIJKE BRILL NV, LEIDEN, 2020 | DOI:10.1163/9789004435292_022

conserve of dog-rose hips, 1 drachm
saffron, 3 grains.
Mix for a dose.

Bernard Penot recommends a water for both kidney stones and bladder stones, for the recipe of which see *Asylum languentium* of the excellent Dr Christoph Petri, our Achilles.

Saffron is also one of the ingredients of Barbette's nephritic spirit. With cold causes and temperaments, nothing is said to be more effective than this. See ibid., *Praxis medica*, p. 209.

Additionally:
Take of: Goose fat, fresh butter,
oils of bitter almonds, scorpions, of each 3 ounces
saffron, 1 drachm; turpentine, 2 ounces.
Mix and apply warm to the kidneys. Elideo de Padoani, in the *Consilia* edited by Scholz, p. 109.

Or
Take of: Oil of wallflower, 1 pound
oils of sweet almonds, sesame, of each 6 ounces
fresh butter, 3 ounces; *saffron* ½ ounce
white coral, ½ pound.
Mix. Make a soft ointment, at the end add: egg yolks, 3 in number. It is effective with pain of the kidneys and of other nervous parts. Stocker, *Empyrica*, p. 124.

For another liniment containing *saffron* see Zecchio, p. 539, *Consultatio* 539; and for Girolamo Capivaccio's ointment and cataplasm with *saffron* see the *Consilia* edited by Scholz, p. 447, *Consilium* 147.

It should be noted that, in cases of bladder stones, anointing in the perineum works excellently and more fruitfully than anointing around the pectineal bone or pubis. In fact, the neck of the bladder is in the perineum, and the stones stick mostly in that very place as it is narrow, and they cause unbearable torments.

When the pain is excessive, sleep (since this is an elixir in medicine, and superior to all gems and precious stones) will need to be induced at all costs with painkillers.

Among the latter, Dr Rivière's anodyne pills excel against nephritis. J.P. Lotichius included their recipe in his *Dissertatio de febribus*, p. 103.

Besides these pills, clysters are not to be disregarded, such as this:

Take of: Roots of marshmallow, of mallow, of each 3 drachms
leaves of mallow, marshmallow,
violets, pellitory of the wall,
flowers of melilot, chamomile,
elder,
tops of dill, of each ½ handful
worms washed in water and sliced, 2 drachms
seeds of anise, dill, of each ½ ounce
linseed, fenugreek, of each 2 drachms.

Boil in cow's milk, 7 ounces. In the colature dissolve:
egg yolks, 2 in number
oils of chamomile, dill,
white lilies, scorpions, of each ½ ounce
saffron, 2 scruples.

Make an clyster.

If the pain is oppressing, you can add Roman philonium to 2 drachms.

You will find another anodyne clyster with *saffron* in my *Tartaro-Mastix Moraviae*, chapter on the stone, p. 231.

11 Urinary Retention

Many have a humorous feeling in common, that, when they hear a sound, immediately they feel the urge to urinate, just like little birds in cages are incited to sing by the rubbing of knives because of the tension of the acute sound, and like the hissing sound of a file causes many to shiver and, from the shivering, externally the hairs are caused to stand on end and internally the muscles jolt. Giulio Cesare Della Scala, in *De subtilitate ad Cardanum* p. 455, believes that this happens because the heat, weak from a minor cause, is agitated, then the sound, sent through the air, moves the air which is inside and everything very close to the muscles. Other things are pulled from their positions by them.

However, this remedy is useless to cure urinary retention, which is the retaining of urine caused by a stone, a lump of blood, pus, or excessive swelling of the bladder, the itch, a tumour or a carbuncle in the bladder, or the ureters, or the urethra. What is urinary retention?

Internally, besides the urinary laudanum of Dr Michael, distinguished for a significant amount of *saffron*, the following water is of benefit:

Take of: *Our anti-colic water of saffron,*
parsley, of each 1 ounce

water of ground-ivy, ½ ounce; volatile salt of amber, 5 grains. Mix.

In the absence of volatile salt of amber, a valid substitute is volatile salt of urine.

Additionally, Feynes, *Practica medicina*, book IV, ch. 4, p. 633, writes that this is particularly good for stimulating urination:

Take of: Greases of rabbits,
hens, of each 1 ounce
galbanum, myrrh,
opopanax or sagapenum melted in white wine, of each ½ ounce; oils of lilies, scorpions, of each 2 ounces; powder of orris, seeds of radish, smallage, of each 1 drachm; *saffron*, ½ drachm; chamomile flowers, ½ drachm. Mix. Make an ointment.

Once it has been besmeared on the pubis and the perineum, the urine flows without delay.

III Impotence in Sexual Intercourse

A woman who is not embarrassed to reveal and expose the secrets of the nuptial bed, and to complain of the impotence of her husband, has always been judged shameless.

Hotman, *Tractatus de dissolutione matrimonii ob frigiditatem*

Pope Honorius imposed a penance on men and women of this sort, who complain of such a thing before three years. Justinian, though, had allowed two years for a husband to prove himself, *Liber penitentialis*, chapter on divorce. With Constantine's *Novels*, 22, however, it was established, and became accepted practice, that it should be three years, and that, after this time, if a wife was found to still be a virgin, and not otherwise, she could separate from her husband.[1] And this not through a divorce as Justinian wanted, but rather through an annulment.

What is impotence in sexual intercourse?

Impotence is the diminished or absent engagement in sexual intercourse, because of spoiled semen, a spell, or a diminution of the spirits, and is caused by cold or too little blood, fasting, or excessive evacuation.

1 It seems Hertodt may be confusing two emperors. Emperor Justinian promulgated Constantine's *Novels* and in novel 22 he amends the previous legislation regarding annulment of marriage due to impotence, increasing the years before which the separation was allowed from two to three.

When impotence originates from coldness of the blood and diminution of the vital spirits:
Take of: Electuary of satyrion roots, 4 ounces
cleaned sweet almonds, pine-nuts, pistachios, of each 1 ounce; condited ginger, condited citron-rind, of each ½ ounce; species diambra, Venice borax, of each 2 drachms; galangal, mace, cloves, long pepper, seeds of rocket, shavings of deer penis, bull, sea-skink, of each 1½ drachms; castoreum, *saffron*, of each 1 scruple; essence of satyrion, 3 drachms; oils of nutmeg by distillation, cinnamon, of each 6 drops; musk, amber, of each 5 grains; syrup of mint, sage, of each ½ ounce. Mix.

Make a strengthening electuary. Take from ½ to 1 ounce of it in the evening before going to bed, and on top of it drink a draught of malmsey wine. Daniel Angelus in *De compositione medicamentorum*, ch. 4, p. 47.

Or
Take of: Pine-nuts, pistachios,
cleaned sweet almonds, of each ½ ounce
seeds of nettle, cinnamon, long pepper, dates, ginger, of each 2 drachms; common hartwort, galangal, of each ½ ounce; seeds of mustard, cardamom, of each 3 drachms; *saffron*, 1 drachm; conserve of satyrion, 1 ounce; whitest sugar dissolved in balm-water, as much as suffices; make an electuary for sexual intercourse. Ferdinando Epifanio, *Empyrica*, p. 405.[2]

The same author, p. 405, describes another electuary with *saffron* against the impotence of kidneys and testicles.

Or
Take of: *Condited saffron bulbs*, 2 ounces
condited satyrion roots, 1 ounce
condited kernels of Indian nuts, condited Chinese ginger, of each ½ ounce; powders of true Indian chocolate, vipers, pearls, of each 2 drachms; fresh pine-nuts, fresh pistachios, of each ½ ounce; oil of nutmeg by distillation, 1 scruple; essence of ambergris, musk, ½ drachm; *quintessence of saffron*, ½ scruple; syrup of cinnamon, spagyrically-prepared, as much as suffices.

Make an electuary. Dose: the quantity of a walnut, before luncheon and dinner with a draught of malmsey wine.

2 Again, the author of *Empyrica* was Nicola Epifanio, not Ferdinando.

Or
Take of: *Oil* or *essence of saffron*, 8 drops or 10
alchemistic gold reduced into gold-leaves and well sweetened, ½ scruple. Mix together, and administer in white Bastardo wine, malmsey or sweet Rhenish wine. Let it be drunk one hour before going to bed. Philip Grüling, *Florilegium chymicum*, p. 524.

The illustrious Rolfinck, *Methodus medicinae specialis*, book IX, ch. 2, p. 823, describes a powder with *saffron*.

Among the medicaments aiding ill-disposed and bewitched nerves, this is of no little importance:
Take of: Castor-beans, 2 scruples
rue, 1 pugil; southernwood, 1 pugil
saffron, 4 grains
whitest sugar, ½ drachm.
Mix. Reduce to powder, and administer with sweet white wine every other day, three hours before meals. Giovanni Zecchio, *Consultationes medicinales*, p. 781.

Secret powder

For the victims of a spell, sorcery, ligature, and those made impotent by witchcraft, there is a secret powder which is made from shavings of elecampane, ox penis, and a little bit of *saffron*.

CHAPTER XX

Diseases of Women

1 Sterility

Alexander the Great, complaining that he did not have a successor for his kingdom, said, "My bereavement is scorned because I have no children." Royal children are the defence of the princely power; and, indeed, a bed without a companion is as miserable as a day without sun, or as a night without stars.

Quintus Curtius Rufus, Juvenal

This happens because of the ineffable and supreme Divine Providence, who does not dispense everything to one single person, but single things to everyone, and, at pleasure, to whom He wishes to. In fact, a marriage does not give children, but GOD, who presides over it, does. You see that a poor man is granted a large progeny and that this is often denied to the rich. The ancients expressed this graciously in the following emblem.

In Athens, there were once two distinguished men, one rich but poor in intelligence, the other extremely poor but rich in intellect. The two Athenians went to Delphi to enquire about the reason for this, and they learned the thoughts of the Oracle through the woman possessed by it.

She gave the wise men two plates of the same shape, workmanship and carving, but not of the same material. In fact, one was made of gold, the other of wood. She gave the first to the rich man, the second to the poor man, and on both plates the following letters were engraved: S.T.L.I.A., with the inscription *Divina Providentia* at the end.

Since the thoughts of the oracle were not clear, they went to Solon, the wisest man in all Greece, who solved the emblem of the rich man in this way:

> *Sufficit Tibi Lamina Ista Aurea*
>
> ["This golden plate is enough for you"]

And that of the poor man:

> *Substantia Tua Lumen Intellectus Animae*
>
> ["The light of the intelligence of your mind is your wealth"]

 | DOI:10.1163/9789004435292_023

And, finally, he added at the end: and this is Divine Providence.

Similarly, even nowadays, the barren, who sow in gravel and sand, can console themselves and follow the practice of Stoicism or of moral education.

What is sterility?

Sterility is impeded conception, caused by the small dimensions of the womb or the excessive humidity of the same, or by sorcery, etc.

The following decoction helps conception:

Take of: Testicles of castrated sheep
prepared in wine and dried,
vulvae of hares identically prepared and dried, mace, ginger, cinnamon, cloves, bishop-weed, of each 2 drachms; *saffron*, 1½ drachms; core or flesh of ordinary nuts, hazelnuts, pistachios, of each 6 drachms. Bruise, macerate and boil at length in 2 pounds of malmsey wine to the consumption of one third of it.

Give 3 or 4 ounces of this decoction to the woman immediately after her normal period, in the morning 3 or 4 hours before taking food, and do this for three days in a row. And on the fourth day she should have intercourse with her husband and, if not entirely barren, she will conceive. Grüling, *Practica medicina*, part 4, p. 437.

Or

Take of: Sweet orange
hollow out a little and fill with *saffron*, 1 drachm, covering with its own lid, and, after placing in a tightly closed jar, put it in the oven when the bread has been taken out and leave it there to roast for a little while. Then pour on it 2 quartarii of Hungarian wine, and let the woman drink it in the morning and in the evening.

11 Chlorosis

Publilius Syrus; Petrarch, *De remediis utriusque fortunae*, book 1, dialogue 2.

Well spoke Sallustius: a very beautiful face is a great richness. In fact, a beautiful face is a silent recommendation. But of all the disappearing qualities of the human body, none is faster than beauty. Socrates said that this is the tyranny of little time.

Combine the beautiful and lively colour of girls, which is transformed into paleness or into chlorosis, and the appearance of the disease, and you will say that it is a mockery of time.

What is chlorosis?

Chlorosis is a change of the natural colour in the face into greenish and pale, with sluggishness, heaviness in all limbs, food aversion, or pica, frequent heart palpitation, or breathing difficulty, sadness, and oedematous tumours especially in the feet and eyelids. It is caused by bad nutrition and abundance of crude humours, which is produced because of the improper disposition of the internal organs, and by the obstruction of the vessels around the womb and of the adjoining organs.

Premised what is to be premised for the preparation of crude humours, Balthasar Timaeus von Güldenklee uses the following medicated wine to remove the obstructions and correct the distemper:

Take of: Leaves of balm, germander,
mugwort, agrimony, of each ½ handful
savin,

roots of elecampane, iris, madder, of each ½ ounce; sarsaparilla, 6 drachms; flowers of wallflower, rosemary, yellow anthers of white lilies, of each 1 pugil; seeds of carrot, anise, fennel, of each 2 drachms; *whole saffron*, ½ drachm. Cut up and shake, pour over them white Rhenish wine, as much as suffices. Drink 1 cup in the morning and in the evening.

III Menstrual Retention

The joy of women is untimely, since, when their menses stop, they immediately believe that they are pregnant and are anxious about the birth.

And this is, but for a moment, like the joy of the hypocrite: although it mounts up to heaven, at the end it dissipates like dung, and those who have seen shall say, where is it? *Book of Job*, 8. 20

Menstrual retention is the stopping of the menses, caused by a fault in the blood or by the paucity of the same because of a previous evacuation, or by the narrowness of the vessels because of a callus, flesh, or a membrane attached to them. What is menstrual retention?

Premised the universal remedies and those that target the thickness of the blood by making it thinner and cutting through it, the following powder is appropriately administered:

Take of: Powder of jalap tinted
with *essence of saffron*, 1 scruple
sulphurated diagrydium, 6 grains
faeculae of bryony, 5 grains
salt of saffron, or Mynsicht's uterine salt, 4 grains
oleosaccharum of savin, 3 grains.

Mix. Make a purging powder.

Or

Take of: Root of round birthwort, the aperient roots,
roots of bryony, madder, of each 2 drachms
black hellebore, 6 drachms
prepared spurge, jalap, of each 3 drachms
agaric, 2 drachms

leaves of mugwort, balm, pennyroyal, of each 2 pugils; flowers of wallflower, borage, marigold, of each 1 pugil; cinnamon, true cassia bark, of each 2 drachms; *eastern saffron*, anthophyllos, of each 1 drachm; whitest salt of tartar, Mynsicht's uterine salt, of each 1 drachm. Mix the cut ingredients in a nodulus made from a white cloth.

We now proceed to the medicaments that put in motion the menses, among which the *essence of saffron* is proposed as a specific by Grüling, *Practica*, p. 448.

The same can be mixed with other ingredients thus:

Take of: Essence of ox-spleen,
mugwort, balm,
liquid iron, of each 2 drachms
savin, *saffron*, of each 1 drachm.

Let it be marked: **Sonderliche essenz** ["Special essence"] 50. 60. drops.

Or

Take of: Compound water of mugwort, 4 ounces
(distilled waters have an extensive penetrative power)
compound syrup of mugwort, 2 ounces
of canella, 1 ounce
essence of balm, mugwort,
liquid iron, ox-spleen, of each 1½ drachms
saffron, 1 drachm.

Mix.

Let it be marked: **Sonderliche Mixtur auff etlichmahl** ["Special mixture for several times"] dose 1. 2. spoonfuls twice a day.

François Joël, book VIII, *Practicae*, section 2, p. 473, describes a decoction for curing menstrual retention, which contains *saffron*.

Or

Take of: Shavings of horse-radish, ½ ounce
root of asarabacca, white agaric, of each ½ drachm

herb of savin, 1 pugil; choice myrrh, 1 scruple
saffron, ½ scruple.

Infuse for one night in white wine, ½ pound. In the morning press them; make a draught which is to be taken in the morning in bed, uninterruptedly for 4 or 5 days before the time of the menses.

For other draughts with *saffron* for inducing the menses, see Feynes, *Practica*, p. 675.

Saffron drives out the menstrual blood to such an extent that Rivière, *Institutiones*, book IV, ch. 22, reports the case of a woman who even incurred a fatal uterine haemorrhage due to the excessive use of *saffron*. Observation on *saffron*

And, therefore, Balthasar Timaeus von Güldenklee did not hesitate to mix *saffron* with his powder to induce menstruation:

Take of: Yellow anthers of white lilies,
flowers of wallflower, marigold,
eastern saffron, of each 1 drachm

the species of troches of myrrh, 2 drachms; prepared white amber, 4 scruples; castoreum, savin, faecula of bryony, of each 1 drachm; oil of amber by distillation, 15 drops. Mix. Make a powder.

Daniel Angelus, in the *Appendix de compositione medicamentorum*, ch. 4, p. 49, describes another powder with *saffron*, which is very effective when thicker blood has become harmful and spoiled.

Van Foreest made a powder with *saffron*, which he names 'ladies' powder':

Take of: Root of white dittany, 3 drachms
zedoary, sweet flag,
cassia bark, cinnamon, mace,
saffron, citron-rind, of each 1 drachm

Indian wood, troches of myrrh, Venice borax, of each 4 scruples; seeds of Roman coriander, 2 drachms; hartwort, fennel, cumin, anise, bay berries, of each 1 drachm. Mix. Make a powder. Dose: 1 drachm; take often.

Nicolai's powder to induce menstruation is also proven:

Take of: Cassia bark,
rind of cassia fistula,
cinnamon, dittany,
saffron, balm, of each 1 drachm.

Mix. Make a powder. Administer three times, in the morning, with a decoction of wine of mugwort. Stocker, *Empyrica*, p. 130.

The most secret of powders

Dr Matthias Franz Hertodt, medical practitioner, kept for himself the most secret of powders so far amongst the arcana, whose recipe is this:
Take of: Calcined ashes of doves' dung, 1 drachm
powder of saffron, ½ drachms.
Mix with an appropriate vehicle.

Or

Extra-ordinary powder

Take of: Vitriolated tartar,
salts of mugwort, amber,
Mynsicht's uterine salt, of each 6 grains
saffron, 4 grains.
Make a powder, which is extraordinary.
Take of: Best aloes, 1½ drachms
prepared myrrh, 2 scruples
Austrian saffron,
prepared white amber, of each 1 scruple.
With potent wine, make a pill-mass, and soften well with some drops of oil of savin by distillation, or oil of amber by distillation.

Or

Take of: Gum sagapenum prepared with vinegar,
white amber, Venice borax, of each 1 scruple
red myrrh, castoreum, of each 15 grains
saffron, 10 grains.
Oil of cloves, 6 drops. Make 40 pills of gold. Dose: 5. Franz de le Boë Sylvius.

Or

Take of: Gum ammoniac dissolved in vinegar,
galbanum, of each 1 drachm
myrrh, borax, pepper, savin, of each 1 scruple
extract of ox-spleen, gentian,
saffron, of each 1 scruple
vitriolated tartar,
crocus martis niger, of each ½ scruple
oil of lavender by distillation, of savin by distillation, of each 6 drops.
With syrup of feverfew, as much as suffices, make pills the size of vetch seeds. Dose: 8 per time in the course of several days. Rolfinck, *Specialis methodus medicinae*, p. 840.

Additionally, before the expected time of menstruation, when you take aperients by mouth, the following will be useful:

Take of: Oil of wallflower, 2 ounces
of *saffron*, 1 ounce
fat of hens, of badgers, of each ½ ounce
saffron, cinnamon, mugwort, of each ½ scruple
wax, as much as suffices.

Make an ointment. With it anoint the lower belly, and the region of the backbone. Feynes, *Practica*, book IV, ch. 57, p. 678.

If fumigation is pleasing:

Take of: Myrrh, calamite storax, of each ½ drachm
white hellebore, madder,
asarabacca, lesser centaury,
leaves of rue, mugwort,
grains of savin, of each 1 drachm
schoenanth, sweet flag,
cinnamon, of each 2 scruples
saffron, 1 scruple.

Make a powder for fumigation. The woman should be covered round about, so that she does not suffer from a headache. Ibid., p. 81.

One may also administer to women the following aperient pessary, which should not be kept long to avoid causing exulceration:

Take of: Essence of the vegetable substances, 2 scruples
extract of saffron, savin,
gum sagapenum, of each 1 scruple.

With honey make a pessary with a string, the length and width of the middle finger.

IV Uterine Haemorrhoids

Themistius said that Prometheus, son of Iapetus, softened the mud out of which he shaped man not with water but with tears. Hence, it is said that pain is part of man. And the pain of the blind haemorrhoids of the womb is not foreign to humankind.

Haemorrhoids of the womb are small swellings similar to the haemorrhoids of the anus, but appearing in the cervix of the uterus, caused by the abundance of feculent thick blood swelling up the openings of the vessels. What are uterine haemorrhoids?

When they cause pain:
Take of: Oils of water-lilies, chamomile, of each 1 ounce
juice of lettuce, 2 ounces
barley-meal, 1 ounce.
Boil and add:
saffron, 1 drachm
egg yolks, 3 in number.
Make a cataplasm.

Or
Take of: Ointment of poplar-buds, oil of roses,
oil of sweet almonds, fresh butter, of each ½ ounce
saffron, 1 scruple
egg yolk, 1 in number.
Make an ointment. When the pain becomes more severe, add some opium. Sennert.

If pain comes with cancer, Feynes makes little pessaries with *saffron*, woman's milk, opium and lanolin. See his *Practica medicina*, ch. 48, p. 657.

v Hysteric Passion

A storm threatens before it arises, buildings crack before they collapse, smoke foreshows a fire.

Seneca

Not differently, with the hysteric passion, a miserable appearance, a sound in the belly like the croaking of ravens and the hissing of snakes, forebodes a forthcoming hysteric storm, no less. Faintness of the body, weakness in the legs as if they were the tottering posts of the human body, and vapours and fumes mounting up from the womb forebode an imminent and directly-following paroxysm.

What is hysteric passion?

Hysteric passion is, in fact, a congeries of symptoms sometimes hurting the natural, animal functions, sometimes the vital functions, by periods, together with a refrigeration of the whole body. It is caused by a malignant vapour from the womb, because of spoiled semen, or, according to the modern medical writers, by corrupted fluids twitching and shaking the womb, and, with it, dragging along the membranes and the nervous parts of the whole body.

With this condition, all medicaments are employed which are good to strengthen the womb, to dissipate malignant vapours, to induce menstruation or to destroy the semen.

But as a doctor always directs his attention to the cause, thus we recommend the following:

Take of: Purified juices of balm, borage, of each 2 pounds
best saffron, 1 drachm

Mix, distil with a water bath. 1 spoon of this water can be mixed into broths. Rivière.

Take of: Extract of castoreum,
essence of saffron, of each ½ scruple
salt of pearls, 6 grains
oil of amber, 2 drops.

Mix and administer with water of mugwort and balm. Hartmann in *Praxis chimiatrica*, p. 181.

Pierre Potier too makes his hysteric extract conspicuous with *saffron*, see his *Pharmacopoea spagyrica*, book I, p. 360. Dose of the same: ½ drachm to 1 drachm.

Marco Gattinara, *Praxis medica*, book IXX, p. 273, prescribes this powder:

Take of: Seeds of chaste tree, carrot,
cinnamon, *saffron*, of each ½ scruple.

Mix. Make a fine powder. Take with water of catmint, mugwort, bugloss, of each 1 ounce.

Or

Take of: Best cinnamon,
seeds of bishop-weed, hartwort, of each 2½ ounces
cumin, 3 ounces; fennel,
smallage, caraway, of each 1½ drachms

nutmeg, sweet flag, galangal, of each 1½ drachms; ginger, mace, cloves, of each 1 ounce; *eastern saffron*, 3½ drachms. Make a fine powder, according to Art. Zaccharias Neucrantz.

Additionally, the illustrious Dr Matthias Franz Hertodt, widely known physician, whom I have always regarded as a father, gave permission to insert in this book a very secret great plaster for the womb so far tried on several magnates, whose recipe is this: The secret plaster

Take of: White wax washed in sweet wine, 3 ounces
or in its place, gum galbanum dissolved in
vinegar, as much as suffices, until it assumes
the consistence of pottage;

extract of saffron,
of castoreum, of each ½ ounce
distilled oils of amber, of coriander, of each 2 drachms
salts of mother-of-pearl,
of red coral, of each 3 drachms
mugwort, tansy, valerian, catmint, of each 1 drachm
English tin, balm,
dittany, of each 1 scruple & 3 grains.
Mix. Boil over the slowest fire to the consistency of a plaster.

Alternatively, infuse this liquor into the womb:
Take of: Water of orange-flowers, 1 pound
basil, avens, 4 ounces; gallia moschata, 2 drachms; galbanum, 3 drachms; musk, amber, of each 5 grains; syrup of mugwort, 3 ounces; *powder of saffron*, cinnamon, cloves, of each 1 scruple. Mix. Mercato, *Praxis medica.*

VI Δυστοκία [Dystocia],[1] or Difficult Birth

Procopius, *De bello Persico*, book I

Someone will probably undertake anything for the sake of children. Often, for the sake of them, a mother is not terrified that the living child is torn away from her whilst she is still alive. And, although the suffering is hard and full of danger, she is not afraid, even if it comes to death, but she dies with calm spirit, as she leaves behind a child. In fact, children are counted among the main gifts of happiness.

What is difficult birth?

Difficult birth is the hindered action of the womb when it has to expel the infant. This happens because of a large infant or a weak infant unable to break the membranes, or simply a dead infant or having a mola attached to it, or because of the weakness of the mother, breathing difficulty, the ducts not being slippery enough, or because of a diminution of the spirits from excessive evacuation or previous illness.

François Joël, for women with difficulties in childbirth, always used with great success the powder noted below, which eases the delivery and draws out the child without harm:
Take of: Borax, 2 scruples; whitest amber, 1 scruple
saffron, ½ scruple.

1 The original heading of this section was incorrect, and used the erroneous spelling, *Διςοκία* [Disokia].

Grind very finely, mix and administer to women in labour with a draught of wine.

Henri de Heer especially recommends this powder:
Take of: Prepared horse testicle,
Venice borax, of each 1 drachm
saffron, 1 scruple.
Administer in water of mugwort distilled with wine.

When a woman is not able to give birth, Jean Prévost's powder is also extraordinary:
Take of: Powder of rind of cassia fistula, ½ ounce
cinnamon, 2 scruples; *saffron*, ½ drachm.
Administer with malmsey wine. Alternatively, boil in wine, strain, and administer in wine. *Selectoria remedia*, p. 108.

Florenzola's powder is also effective, which is this:
Take of: Choice cinnamon, white dittany,
cassia bark, rind of cassia fistula,
saffron, balm, of each ½ drachm.
Mix, make a fine powder, and administer three times in broth of red chick-peas.

Or
Take of: Savin, 2 scruples
myrrh, borax, of each ½ scruple
cinnamon, 1 scruple; *saffron*, 10 grains.
Make a powder. Administer with water of white lilies or, better, with sweet wine or malmsey; additionally, anoint the belly with *oil of saffron* or white lilies and you will see an extraordinary result even with patients in danger of death.

For us the following powder is a specific: Specific powder
Take of: Eel's liver dried in a bread oven, 1 scruple
peach-kernels, 7 in number
kernels of walnuts, 7 in number
saffron, 7 grains.
Administer with water or decoction of larkspur, or other convenient vehicle, such as water of mugwort, or of white lilies.

Saffron is included also in Dornkreills' powder against difficult birth, as well as in that of Elideo de Padoani, never sufficiently recommended.

Or

Take of: Cassia, 12 grains
white dittany, ½ scruple
white amber, 8 grains

Venice borax, 1 scruple; seeds of lavender, 4 grains; red myrrh, 6 grains; *saffron*, ½ scruple. Mix.

Let it be marked: Nothpulver ["Emergency powder"]. Administer in the following water:

Take of: Water of white lilies, ½ ounce
pennyroyal, mugwort, of each 3 drachms
tincture of cassia, 2 drachms. Mix.

Amato Lusitano, *Curationum medicinalium, centuria* 5, p. 71, recounts that, from repeatedly drinking the medicament that I am going to describe, a woman gave birth to two completely yellow girls.

The author claims that such pigmentation was caused by *saffron*, as, once washed, they were both white.

And we will ascertain whether this is possible in the questions that we are going to discuss at the end of this book. This is the medicament:

Take of: Best cinnamon, troches of myrrh,
best saffron, of each ½ drachm.

Grind everything extremely well and finely, and give to drink in potent wine.

Take of: Flowers of wallflower, Venice borax,
prepared white amber,
shavings of savin-wood,
gallia moschata,
eastern saffron, of each ½ scruple

confection of kermes-berries, ½ drachm; prepared pearls, ½ scruple. Mix with a little water of cinnamon and drink immediately.

Many women with François Moreau of Brussels noticed that, among all medicaments inducing childbirth, this one not only deserved the first place for facilitating the delivery, but it was also an excellent remedy for retained secundines, or to expel a dead foetus. See his newly published work *De febre maligna*, p. 171.

Or

Take of: Borax, 1 drachm or 4 scruples
cinnamon, *saffron*, of each 1 scruple
ginger, ½ scruple.

With confection of kermes-berries (which, besides strengthening, awakens an inactive faculty), make a bolus. Felix Platter, *Praxis medica*, p. 463.

Sennert, book III, *Practica*, p. 417, claims that childbirth is particularly promoted by the following troches:

Take of: Secundines and umbilical cord
dried in an oven and powdered, 2 drachms
cinnamon, 1 drachm; *saffron*, ½ scruple.

With juice of savin, make troches. Administer 2 drachms in a suitable vehicle.

Or

Take of: Spirits of secundines,
saffron, of each 2 drachms.

Mix. Dose: 30–40 drops in a convenient vehicle.

Or

Take of: Oils of amber, pennyroyal,
saffron, of each 2 drops
sugar, ½ scruple. Make a mixture.

Administer with waters of pennyroyal, and white lilies, of each ½ ounce. It is excellent.

Alternatively, at intervals, administer to women with difficulties during labour 3 ounces of the following decoction:

Take of: The aperient roots, 1 ounce

betony, mugwort, wild marjoram, feverfew, pennyroyal, horehound, of each 1 handful; seeds of anise, smallage, fennel, of each 2 drachms. Boil to half a pound, and in the strained juice dissolve sugar, as much as suffices; cinnamon, 2 drachms; *saffron*, ½ drachm. Feynes, *Praxis medica*, p. 722.

Or

Take of: Roots of dittany, ½ ounce
of round birthwort,
madder,

rind of cassia fistula, of each 2 drachms; herb of mugwort, pennyroyal, savin, of each 1 handful; flowers of lavender, chamomile, wallflower, of each 1 pugil; cinnamon, 3 drachms; *saffron*, 1 drachm. Boil in white wine, drink the draught from this. Felix Platter, *Praxis medica*, p. 462.

Take of: Compound water of mugwort, 3 ounces
essences of mugwort, of balm, of each 1 drachm
saffron (which is unparalleled), 2 scruples
compound syrup of mugwort,
of canella, of each 3 drachms.
Mix for two doses.

Or
Take of: Waters of balm, mugwort,
cinnamon, the hysteric water, of each 1 ounce
choice myrrh, ½ drachm
prepared amber, 1 scruple; *saffron*, ½ scruple.
Make a small draught to be drunk, warm, during delivery. Balthasar Timaeus von Güldenklee. *Casus medicinales*, p. 223.

Or
Take of: Venice borax, ½ drachm
saffron, prepared white amber, of each ½ scruple
water of mugwort, syrup of mint, of each 1 ounce.
Mix. Werner Rolfinck, *Methodus medicinae*, p. 840.

Additionally, this uterine pessary can be prepared:
Take of: Galbanum dissolved in vinegar, 1 ounce
myrrh, 2 drachms; *saffron*, 1 drachm.
With oil of orris make a pessary. Sennert.

Alternatively, anoint the navel, pecten, and private parts with the following liniment:
Take of: *Oils of saffron by insolation*, wallflower,
white lilies, of each ½ ounce
extract of saffron, 1 scruple
grease of snakes, 2 drachms; a little bit of wax. Make a liniment, with this precaution: when the infant is brought forth, carefully wipe off this liniment for fear of a more severe haemorrhage.

VII Pain after Childbirth

We persevere with unsuccessful experiments to such a degree that, conquered, we go again to war, shipwrecked, we go again to sea.

The offspring about to come causes the oblivion of the pain one has gone through: in fact, nothing is dearer to humankind than being granted by nature, children.

Cicero, *Oratio ad Quirites*

The pain after childbirth is an unpleasant feeling in the womb, caused by grumous blood or by lochia not flowing properly or by constant looseness.

What is the pain after child-birth?

With this pain, the following medicament is proven by Helwig:

Take of: *Saffron*, ½ drachm; mace, 1 scruple
tryphera magna, ½ drachm.
Mix for two doses. Administer in warm wine.

Or
Take of: Seeds of hartwort, cumin, zedoary,
galangal, *saffron*, castoreum, of each 1 scruple
oil of cinnamon, 5 drops.
Mix. Make a powder for three doses in pennyroyal-water and Rhenish white wine mixed together, and take with egg yolk, mace and a little sugar in the form of a broth. Balthasar Timaeus von Güldenklee, *Casus medicinales*, p. 229.

Or
Take of: Seeds of hartwort, 2 drachms
cumin, dill, caraway, of each ½ scruple
rind of orange, roots of zedoary,
galangal, of each 2 scruples; *saffron*, ½ drachm.
Mix. Make a powder. Give in a box. Dose: 2 or 3 points in broth.[2]

When the pain comes from grumous blood, we have tested this broth many times:
Take of: Roots of couch grass, ½ ounce
herb of chervil, 1 handful
Boil with spring water in the belly of a chicken, after adding *saffron* and mace at pleasure.

2 Points: meaning here the amount of powder that will sit on the point of a knife.

Or
Take of: Lovage-water, ½ ounce
spirit of saffron, ½ scruple.
Mix.

VIII Mola

Often, for a fault in the parents, children are not only born blind, lame, maimed or sickly in some other way, but also ugly, stupid, and disposed to evil. Indeed Johann Remmelin, in *Catoptrum microcosmicum*, is witness that, by leave of GOD, because of lust, undoubtedly unseemly intercourse, perverse imagination and such like, unnatural offspring, by all means monsters and abominations, are generated.

What is a mola?

A mola is a mass without bones and organs produced from incomplete conception because of a fault in the seed sometimes of the man, sometimes of the woman.

This powder is excellent to expel it:
Take of: Powder of troches of myrrh, 6 drachms
asafoetida, 1 drachm; dittany of Crete,
powder of savin, of each 2 drachms
prepared white amber, 1 drachm; mineral borax, 2 drachms; cinnamon, cassia, of each 1½ drachms; *eastern saffron*, ½ drachm. Make a powder. Dose: 1 or 2 drachms in decoction of red chickpeas with parsley-root.

Or
Take of: Choice cinnamon, long birthwort, of each 1 drachm
leaves of savin, 2 drachms
root of asarabacca, *saffron*,
gum lac, of each ½ drachm
seeds of rue, 2 scruples.
Mix the extremely finely ground ingredients. Drink 1 drachm of this in the morning and in the evening with a draught of Rhenish wine. Joël, book VIII, section 3, p. 484.

IX Retained Secundines

When nature fails in its work, it is necessary to employ evacuation, and nature's failure must be remedied by science. Hippocrates, 2, *Aphorisms* 20 and 22.

The secundines solicit their natural expulsion, and since, if they are retained, very severe symptoms can follow, even death, one must try in every way to expel them as fast as may be.

The secundines are, however, sometimes retained by nature. They are the membranes surrounding the foetus, retained because of a strong connection with the bottom of the womb, or, if they have already separated from the womb, because of them being convoluted like a ball, or because of the closing of the mouth of the womb.

What are retained secundines?

The following powder is proven to expel them:

Take of: Choice cinnamon, shavings of ivory,
saffron, of each 1 drachm.
Administer in sweet wine.

Tulp, *Observationes medicinae*, book IV, ch. 41, p. 363, saw the following work on several women in childbed:

Take of: Venice borax, white amber, of each 1 scruple
saffron, cassia,
castoreum, of each ½ scruple.
Mix. Make a powder.

Franz de le Boë Sylvius commends this one:

Take of: Red myrrh, 1 scruple
Austrian saffron, ½ scruple
Venice borax, ½ drachm
Ceylon cinnamon, 8 grains.
Mix. Make a powder, divide into two doses.

Balthasar Timaeus von Güldenklee prescribes this powder:

Take of: Shavings of date stones,
troches of myrrh,
white amber, *saffron*, of each 1½ scruples.
Mix. Make a powder for six doses. Take with warm wine in the morning, evening and afternoon.

Stocker in *Empyrica*, p. 127, prescribes this:

Take of: Mithridate,
saffron, of each 1 drachm.
Administer with wine.

A decoction of schoenanth, taken after *saffron* has been added to it, expels the secundines and soothes the pain. Luigi Bonacciouli, *De formatione foetus*, p. 312.

x Retained Lochia

Among the principles of political life, which age has established with great authority, this is not the least: old laws must not be changed without purpose.

The law of nature sanctions that the lochia and secundines must be duly evacuated after childbirth. If this not observed, if this is changed without purpose, or if these things are evacuated only a little or hold back altogether, they are the cause of many illnesses. Galen, 1, 3 *Commentary on Hippocrates' epidemics*, text 73. And an account of this is in Hippocrates in the work just quoted.

What is retained lochia?

Retained lochia is the retention after childbirth of the blood collected during the whole time of gestation in the vessels disseminated around the womb.

In this instance, the powder described in Philip Grüling's *Praxis medica*, book IV, part 4, p. 453, is very effective:

Take of: Anthophyllos, choice cinnamon,
nutmeg, mace,
saffron, savin, prepared pearls, of each as much as you will.
Mix. Dose: 1 drachm in wine.

When, with the retention of the lochia, a severe disorder of the blood with vomiting, thirst and wakefulness appears, the illustrious Dr Thomas Willis recognized that laudanum mixed with *saffron* is often employed with great success, and also the following mixture of cooling julep:

Take of: Waters of pennyroyal, balm, of each 3 ounces
the hysteric water, 2 ounces
syrup of mugwort, 1½ ounces
tincture of saffron, 2 drachms
castoreum, bound in a patch and suspended, 1 scruple.
Mix. Administer 3 or 4 spoonfuls of this several times a day. See the well-researched tract *De febribus*, p. 285.

CHAPTER XXI

Fevers

Thirst is the most pressing of symptoms, and it troubles especially those suffering from fever. Hieroglyphics represent beautifully a man of such kind, consumed by a fiery thirst, as someone whom the Dipsas bit, as the bite of such a snake inflames to such an extent that the persons bitten look as if they are just about to catch fire and as if they are consumed by fire. Once, a column marked with this hieroglyphic character, in which a dying man had been hidden, whose feet the snake bit with his teeth and swiftly entangled in his coils as if laid around ostrich eggs (which this serpent especially watched for), was a sight on the shore of the Greater Syrtis between Libya and Egypt.[1]

Fever is, in fact, a condition, when, from the irritation of the heart originating from the disproportioned fermentation of humours, one finds oneself with frequent and irregular rapid pulse, natural heat more or less increased and thirst among the things that overtake health. What is fever? Franciscus Piens

Premised what is to be premised, Platter's anti-pyretic powder, which *saffron* improves, is excellent:

Take of: Powders of river crabs,
human skull, shavings of ivory, of each 1 scruple; cinnamon, ½ drachm; *saffron*, 1 scruple. Make a powder. Dose: 1 drachm with wine. See his *Praxis*, p. 166, as well as *Gammarologia* by Dr Sachs von Löwenheim, p. 721.

In cases of quartan ague, the following plaster was a panacea for the Prince of Anhalt, as it drew out the miasma of the fever to such a degree that it immediately even produced pustules:

Take of: Pepper, salt,
eastern saffron,
garlic that is covered by earth, of each, as much as suffices.

Mix. Bruise in a mortar to make a cataplasm, then [place] in a nodulus, and apply to the external part of the left ring finger. Remove within the same hour in which it has been applied and repeat before a paroxysm.

1 Greater Syrtis: the Gulf of Sidra, on the northern coast of Libya.

 | DOI:10.1163/9789004435292_024

Or
Take of: Spider's web,
yeast, elder,
vinegar of roses, *saffron*.
Mix together, apply to the wrists and heart. Some, like Alexander of Tralles, book I, ch. 15, Sextus and Dioscorides, say that in the stomach of swallows there are little stones which, according to Marcellus, are good against quartan agues if wrapped in a *linen* or *cloth of saffron* and hung around the neck with a thread.

In case of quotidian ague, Hendrik de Roy, *Consilia medica*, prescribes this bolus:
Take of: Treacle, 2 scruples
extract of blessed-thistle, 1 scruple
eastern saffron, 4 grains.
Mix. Make a sudorific bolus.

In case of syncopal fever, when this originates because of a very thin body, Van Foreest applies the following epithem:
Take of: Water of roses, 8 ounces
of sorrel, water-lilies, bugloss, of each 3 ounces
all the kinds of sanders,
both kinds of coral, of each 1½ drachms
saffron, ½ scruple
best white wine, ½ ounce
camphor, 3 grains.
Make a cordial epithem.

In cases of malignant and petechial fever, except for water-pepper, which is a specific for this fever, the following powder is worthy of praise:
Take of: Scorzonera, water-germander,
burnt prepared hartshorn,
sal-prunella, of each 1 ounce
camphor, 2 scruples & 5 grains
saffron, 1 scruple.
Mix. Make a powder. Dose: ½ drachm or 2 scruples.

The illustrious Rolfinck, *Methodus medicinae specialis*, book IX, ch. 1, describes another powder with *saffron*.

To regulate the heat and strengthen the heart, Brunner, in *Consilium* 43, p. 246, adds *saffron* to his ointment and cordial water for external use, or applies the following cordial epithem:

Take of: The cooling powder of pearls, 2 drachms
grains of kermes, prepared coral, of each 1 drachm
eastern saffron, ½ drachm

waters of balm, borage, sorrel, of each 1 pound; Cretan wine, ½ pound. Mix. Pierre Potier, *Observationum et curationum, centuria* II, p. 169.

In cases of English sweating sickness, Frederigo da Fonseca adds *saffron* to his preserving pills too. Fabry von Hilden, *centuria* VI, observation 77, noted a *saffron*-coloured sweating sickness [*sudor croceus*].

Use this powder to cure smallpox:

Take of: Seeds of turnip, 4 ounces

burnt hartshorn, 4 ounces; seeds of melon, 5 ounces; blessed-thistle, citron, of each 3 ounces; charcoal of lime-wood, 2 ounces; myrrh, 1 drachm; safflower, 1 ounce; *saffron*, camphor, of each ½ drachm; laminated gold leaves, 10 in number. Mix. Make a powder. Give in hemp-water.

This decoction contributes to the exsiccation of the pustules of the pox:

Take of: River water, 2 pounds
salt, ½ ounce; lupin, barley, of each 1 pugil
saffron, 1 drachm.

Boil, and touch the pustules with a cotton soaked in such water. Hafenreffer, *De cutis affectibus*, ch. 4, book II, p. 308.

When lungs affected with smallpox cause difficult breathing, and threaten death, the following is good if unnatural heat is absent:

Take of: Zedoary, liquorice, of each 1 drachm
saffron, ½ scruple
gum ammoniac washed and
percolated with juniper-water, 2 scruples.

Mix with syrup of hyssop and frequently assume by mouth and swallow. François Joël, book IX, section 5, p. 520.

The illustrious Dr Marek Marci, once emeritus Professor and practitioner in Prague, in cases of smallpox where catarrh supervened, successfully administered syrup of violets, ½ ounce or 1 ounce with 30 drops of *spirit of saffron*, and repeated this rather often.

Frequently, he also used the same spirit, up to 30 drops, in which 3 grains of camphor were dissolved, with a vehicle, in cases of fevers.

To preserve the whole face from smallpox, wash the face three times a day with rose-water in which some *saffron* has been dissolved.

To preserve the eyes in cases of smallpox, we use the same rose-water well dyed with *saffron*. Grüling, *Praxis medica*, book I, ch. 10.

Or

Take of: Waters of plantain, rose, of each 1 ounce
prepared emeralds, 12 grains
prepared pearls, 6 grains
saffron, 4 grains
camphor, 3 grains.
Mix for a collyrium.

Or

Take of: Powder of sumach leaves, 1 drachm
prepared seeds of coriander or myrtillus, ½ drachm
saffron, 1 scruple; camphor, ½ scruple
water of roses or plantain, 8 ounces.
Mix. Grüling, *Praxis medica*, p. 114.

Or

Take of: Water of plantain, larkspur,
cornflower, egg-white, fennel, of each 1 ounce
roses, ½ ounce
seeds of sumach, 1 drachm; rind of pomegranate, ½ drachm; flowers of red roses, balaustines, of each 1 pugil. Let them boil a first and a second time, strain, add:
sief album without opium, 1 drachm
saffron, 6 grains
camphor, 3 grains
Mix. Instil into the eyes lukewarm, and from the outside moisten the eyelids with a small linen cloth.

Plague

The scourges of humankind

Besides famine and war, plague is one of the scourges with which the wickedness of men is punished by the Ruler of all things. Plague is a deadly contagion, during which the burials of parents are often mixed with burials of children, and often, I may say, one household is nothing other than one burial.

To these three scourges, GOD added a fourth one for the rich, that is, altercation, with which the rich are afflicted and through which they are deprived of their riches. They are diminished to the condition of poor people with extreme misery, as, unfortunately, when they are rich, they grow richer mostly from the hunger of others. In war the Parthians usually always triumphed imitating the plague and can easily escape fighting against it by running away.

This is a contagious epidemic disease, extremely dangerous for humankind. It is caused by a poisonous miasma caught first from the air, then spread through contagion, inflicting a considerable heart irritation with symptoms more severe than those that can be understood from the causes. What is the plague?

The following sudorific of Paracelsus, remarkable with all fevers, especially with plague, fights most eagerly against this condition:

Take of: Purified sulphur, that is flowers of sulphur, 10 ounces
myrrh, 1½ ounces; hepatic aloes, 1 ounce
saffron, ½ ounce
Mix. Make a powder, take with vinegar of squill or rue.

Johann Böckel, physician in Hamburg, in his *Regimen pestis*, p. 86, describes another famous powder with *saffron*.

Caesar's powder against the plague is this:

Take of: The species of Venice treacle,
best mithridate, of each 6 drachms
bole armeniac, corrected sulphur, hartshorn, of each ½ ounce; zedoary, ginger, of each 3 drachms; dittany of Crete, tormentil, burnet, gentian, of each 2 drachms & 2 scruples; camphor, 4 scruples; myrrh, *saffron*, of each 1 drachm; bone of hart's-heart, 2 scruples. Mix. Make a powder. Sprinkle a little bit with brandy. Thomas Jordan, *De pestis phaenomena*, p. 515.

Jodocus Willich commends *saffron* in his expulsive powder, expulsive round tablets as well as in his expulsive mass. See his *Consilia medica*, p. 644 et seq.

Also, the Faculty of Medicine of Marburg put *saffron* in a curative powder prescribed in 1597. See Christoph Petri, *Carduus benedictus*, p. 174.

True animal bezoar, or Mutzler's genuine curative powder, known to few, which the author does not mention in his book and which *saffron* adorns as well, is superior to all of these remedies.

And, in fact, *saffron*, by expelling the pestilential venom from the heart, is of benefit in cases of plague. Hieronymus Bock, p. 286.

Gans, in *Corallorum historia*, p. 154, describes some compound flowers of sulphur against the plague.

Avicenna prepares his preserving pills in the following way:
Take of: Aloes, 3 parts
myrrh, 2 parts
saffron, 1 part.
With potent wine make pills.

Isaac Roet in *Pestis adumbrata*, book III, *Aphorism* 32, p. 70, prepares Ruffus' pills, most excellent in cases of plague, in the following way:
Take of: Rosated aloes, 2 ounces
gum ammoniac in drops, 2 ounces
myrrh, 1 ounce; *saffron*, ½ ounce.
Pour on spirit of wine until a tincture is extracted, then let the tincture evaporate as much as suffices to make a mass for the pills.

Crato prescribes these anti-pestilential pills:
Take of: Rosated aloes, 1 ounce
myrrh, *saffron*, of each ½ ounce
troches of agaric, 2 drachms; choice rhubarb, 1 drachm; bole armeniac, 2 scruples; aloes-wood, white dittany, prepared pearls, of each 1 scruple. With lemon juice, make small pills. Dose: 1 drachm every week.

Other pills can be produced in the following way:
Take of: *Saffron*, myrrh, camphor,
hartshorn, spodium, of each 1 drachm
aloes-wood, white behen, of each ½ drachm; terra sigillata, ginger, of each 2 scruples; ambergris, 1 scruple; fragments of hyacinth, emerald, garnet well prepared, of each ½ drachm; choice agaric, choice rhubarb, of each ½ ounce; aloes from a bladder to the equivalent weight of all the ingredients.

With syrup of lemon, make a pill-mass. Dose: ½ drachm shortly before meals.

Timaeus von Güldenklee describes other pills against the plague, containing *saffron*, [see] *Epistolae medicinales*, p. 326.

Additionally, *saffron* is added to the anti-pestilential pills prescribed by the most distinguished Faculty of Medicine in the famous Academia Gissena in the year 1626.[2] See Gregor Horst, part 2, book VII, observation 29; Andreas Graba, *Elaphographia*, p. 237.

2 The Academia Gissena or Gissennsis was located in the German town of Giessen.

Mutzler, in his *Bezoardicum animale*, describes some pills not unworthy of notice, in which he mixed *saffron* as well, p. 157.

Saffron finds its way into Untzner's syrup against the plague, *Antidotarium*, book II.[3]

Moreover, the antidote of scorzonera of King Philip II of Spain honours *saffron*. See *Scorzonera curiosa* by our most esteemed President, p. 140.

And *saffron* did not want to be absent in the Spanish antidote, whose recipe is this:

Take of: Venice treacle, 3 ounces

Venice mithridate, 1½ ounces; Maximilian's electuary of eggs; Fracastoro's diascordium, of each ½ ounce; extract of gentian, angelica, of each 1 drachm & 15 grains; black hellebore, 1 drachm; *oil of saffron*, ½ ounce; oils of citron, rosemary, cloves, white amber, rue, camphor, nutmeg, juniper, of each 1 drachm; balsam of sulphur, 1 scruple; oriental terra sigillata, 1½ drachms; prepared bole armeniac, 1 drachm; soluble magistery of pearls, hartshorn, of each 2 scruples & 5 grains; the species de gemmis frigidi, diambra, diatrionsantalon, of hyacinth, liberans, of each 2½ scruples; oriental bezoar, 1½ drachms; confection of kermes-berries, 1½ drachms; roots of angelica, sweet flag, round birthwort, white dittany, gentian, greater celandine, burnet, butterbur, valerian, zedoary, of each 2 scruples & 5 grains. With syrup of sorrel, citron, as much as suffices, make an electuary. Dose: ½ drachm or 2 scruples to 1 drachm. For preservation, [make pills] the size of juniper-berries or peas.

Once, in Moravia, a proven electuary for health was this:

Take of: Best aloes, terra sigillata,
bole armeniac, of each 1 ounce
root of white dittany, carline thistle,
angelica, choice myrrh, of each ½ ounce

camphor, gum ammoniac, *Austrian saffron*, choice rhubarb, of each 1 drachm; Venice treacle, ½ ounce; despumed honey, as much as suffices. Make an electuary.

Those who wish to have tablets should use Ferdinand I's tablets against the plague, whose recipe is found in Wittich, *Consilia Germanica*, p. 22, as well as in our illustrious President's *Scorzonera curiosa*, p. 64.

For those who prefer round tablets:

Take of: Extract of elecampane, zedoary,

3 Untzner is misspelled by omission of the letter t in the original version of *Crocologia*.

angelica, burnet, of each 1 scruple
saffron, ½ scruple
cream of sulphur, 1½ scruples; magistery of hartshorn, ½ scruple; salt of coral, pearls, of each 1 scruple; true bezoar stone, ½ drachm; distilled oils of angelica, amber, citron, of each 4 drops; sugar, 4 ounces. Dissolve in balm-water. Make round tablets. Andreas Spenholz, *De peste*, p. 99.

When sufficient evacuation through the skin has already happened then these round tablets can also be administered, which fight the malignity and quench the burning sensation in the inner parts of the body:
Take of: Solutio perlarum,
solutio margaritarum, of each 1 scruple[4]
hyacinth,
extract of saffron, ½ scruple
confection of kermes-berries, 1 scruple; oil of citron-rind, 4 drops; whitest sugar, dissolved in rose-water, as much as suffices. Make round tablets. Isaac Roet, *De pestis adumbrata*, book V, *Aphorism* 28, p. 84.

The famous Rolfinck in his treacle water, with every great praise, certainly does not neglect *saffron*. See his *Methodus medicinae specialis*, book XI, ch. 11, p. 914, as well as Tentzel's most excellent pestilential red water.[5] See his tract on the plague.

Those who like mixtures:
Take of: Treacle, 1½ ounces
camphor, 2 drachms; *saffron*, 1 drachm.
With aqua solis, make a mixture. Jordan, *De pestis phaenomena*, p. 527.

In *Chymia in artis formam redacta*, p. 267, Rolfinck prepares this mixture against the plague:
Take of: The simple diaphoretic mixture, 4 ounces
essence of juniper grains prepared
with malmsey,
essences of angelica, rue, water-germander, balm, blessed-thistle, gentian, of each 1 ounce; *saffron*, 2 drachms; spirit of hartshorn with volatile salt, 1

4 *Perla* and *margarita* are synonyms for "pearl", but Hertodt's usage here implies they are different ingredients, unless the word "or" has been omitted accidentally.

5 Tentzel is misspelled by omission of the middle letter t in the original version of *Crocologia*.

ounce; rectified spirit of guaiacum, ebony, of each 2 drachms; syrups of sorrel, citron, water-germander, blessed thistle, cherries, cloves, scorzonera, of each 5 ounces. Let it digest for 24 hours; keep in a vessel in which a nodulus of ambergris, ½ scruple, musk, 1 scruple, has been suspended. It is a famous alexiteric. Dose: 2 to 3 drachms in wine as a preservative, in water of water-germander as a cure. Our most esteemed colleague Christoph Petri, *De carduo benedicto*, p. 190.

And *saffron* is not the least important ingredient in the essence of bezoar of a certain doctor very much known across the whole of Europe, for whose recipe see *Scorzonera curiosa* by our most esteemed President, Michael Fehr, p. 145.

Tincture of bezoar widely proven by experience

We present below another tincture of bezoar, which is widely proved against the plague:

Take of: Powder of freshly-prepared troches of vipers, 4 ounces
powder of bone of vipers, ½ ounce
rob of juniper, 1 ounce
saffron, 4 scruples

pour into rectified spirit of wine, 2 pounds; spirit of urine of a child drinking wine, 1 pound. Leave to digest in a closed vessel until they acquire a golden or a very red colouration. Store the colature for use in a well protected glass.

Some drops of this tincture in wine, or in a suitable vehicle, preserve the heart from poison, dissipate the malignant matter through sweat in a greater quantity, and are an elixir against all sorts of poisons and malignant diseases.

With this very effective tincture, Dr Matthias Franz Hertodt, medical practitioner and physician in Nikolsburg, in the year 1665 (since at the time there was an outbreak of plague there) protected himself, his relatives and myself, and many others.[6]

Elixir for strengthening and preservation. An elixir of such kind in cases of plague can be the *elixir of citron and saffron*:

Take of: Fresh citron-rind, the greatest part
of the internal white pulp having been removed, 4 pounds.

Cut into small pieces, afterwards infuse in 4 measures of aqua vitae for 24 hours, then, by means of a bladder, distil all the spirit of the infusion (which

6 Nikolsburg is now called Mikulov, a Moravian town in the present-day Czech Republic.

will be perhaps the half part). You will know that all the spirit has gone through if paper moistened in the passing spirit takes fire easily, or with difficulty.[7]

Tinge the very fragrant distilled spirit with *best saffron*, ½ drachm, mixed with 2 pounds of syrup of citron juice.

Dose: 1 spoonful in the morning.

Additionally, epithems can be employed for strengthening. In fact, they are not astringents or repellents, but they foster the heart's own heat.

Among the remedies of such kind, this is good:

Take of: Waters of bugloss, sorrel, of each 5 ounces
red sanders, 4 scruples
orange-rind, 2 scruples
saffron, ½ scruple.

Mix. Make an epithem for the heart. Isaac Roet, *De peste adumbrata*, p. 33.

Or

Take of: Rose-water, 4 ounces
waters of balm, borage, bugloss,
sorrel, of each 2 ounces
vinegar of roses, 2 ounces
powders of flowers of roses, both corals,
white and red sanders, of each 1½ drachms
saffron, 2 drachms; camphor, 2 drachms.

Mix. In this, soak cloths made fourfold and then squeezed. Apply lukewarm to the heart. Rolando Capelluti, *Tractatus chirurgicus de peste*, p. 115.

Alternatively, apply the following cordial sachet:

Take of: Roots of valerian, plantain,
burnet, angelica, of each 3 drachms

flowers of clove carnations, red roses, of each 2 handfuls; herb of balm, basil, rosemary, 3½ handfuls; yellow sanders, 4½ drachms; amber, 3 drachms; calamite storax, 6 drachms; nutmeg, zedoary, cinnamon, citron-rind, of each 3 drachms; *whole saffron*, 2 drachms; camphor, ½ drachm; oil of angelica, 12 drops. Cut up, shake, place in sachets.

For Girolamo Capivaccio's cordial sachet with *saffron* see Christoph Petri, *Asylum languentium*, p. 263.

7 This simple test works because of the alcoholic spirits' flammability.

Alternatively, anoint the heart with this liniment:
Take of: Best treacle, ½ drachm
juice of lemons or citrons, ½ ounce
vinegar of roses, 2 drachms; confection of kermes-berries, ½ drachm; camphor, 2 scruples; *saffron*, ½ scruple; musk, 4 grains. Make a liniment.

Or
Take of: Ointment of yellow water-lilies, 1½ drachms
Venice treacle,
best mithridate, of each ½ ounce
clarified juice of borage, 3 drachms; oil of mace, ½ drachm; *saffron*, 2 drachms; species cordiales temperatae, the species for cordial epithems, of each 1 drachm. Mix. Make a liniment. Isaac Roet, *De peste*.

For a perfume, make this sachet:
Take of: Roses of Provence,
white and red sanders,
leaves of bay, valerian, avens, cloves, cardamom, nutmeg, spikenard, grains of paradise, storax, of each 2 drachms; camphor, 2½ drachms; *saffron*, 1 drachm; musk, 1 scruple. Make all into a powder. Give as a sachet. Rolando Capelluti, *Tractatus chirurgicus de peste*, p. 115.

Paracelsus in his xenexdo, which the famous Thomas Bartholin describes in *Cista medica* and Grüling in *Praxis medica*, book v, part 2, p. 522, does not exclude *saffron*.[8]

Besides the internal remedies listed above, the following is useful on the outside to avoid the virulent ulcers and carbuncles spreading:
Take of: Venice treacle,
mithridate, Venice soap, of each ½ ounce
honey of roses, 1 ounce; crushed salt, 1 drachm; soot, 2 ounces; *saffron*, ½ ounce; egg yolks, 30 in number. Mix and make a liniment; apply to the ulcers as a plaster. Karl Widemann, *Von der pest*, p. 129.

To make the buboes soft and ripe:
Take of: Venice treacle, mithridate,

8 The reference quoted by Hertodt, Grüling (1665), p. 521, gives a recipe for zenechdon, and on p. 522 two different versions of diazenech; all of which contain saffron and arsenic amongst other ingredients. We therefore assume Hertodt's word "xenexdo" is synonymous with zenechdon.

Venice soap, of each ½ ounce
crushed salt, 1 drachm; *saffron*, 2 drachms,
egg whites, 3 in number.

Make an ointment, according to Art. Smear on the buboes, ibid., p. 128.

For a maturative:

Take of: Corn-meal, 3 handfuls[9]
scabious-water, oil of chamomile, as much as suffices.

Boil, add:
saffron, 1 scruple. Mix.

As a mundificative for an aposteme, Crato administers this:

Take of: Turpentine washed with scabious-water, 1 ounce
juice of scabious, 2 ounces
egg yolk, 1 in number
saffron, ½ scruple.

Mix. Make an ointment.

9 Meaning wheat or barley meal, not maize meal.

CHAPTER XXII

Diseases of the External Parts

1 Erysipelas

Although patients consider erysipelas less terrible than other diseases, it is, however, bad if mortification or suppuration accompany it, says Hippocrates, *Aphorism* 20, section 7. In the same way, the same author also claims that if erysipelas seizes a pregnant woman, it is fatal.

Erysipelas is a slight swelling spread out across the skin, caused by thin bilious blood or rather by moist sourness molesting the archeus of the part and growing hot there. What is erysipelas?

Our syrup of saffron is very good for soothing the archeus internally.

Externally, besides the fumigation mentioned in the chapter on the uses of preparations of *saffron*, a single *saffron-flower* laid on the affected part benefits, Bartholomaeus Anglicus, *De proprietatibus rerum*, book XVII, ch. 41, as does Mynsicht's *cloth of saffron*:

Take of: Mill-dust, 3 ounces
seeds of henbane, 2 ounces
ceruse, litharge, of each 1 ounce; *saffron*, 2 drachms.

Mix. Make a powder. Apply to the affected part with hemp tow. Daniel Angelus, *De compositione medicamentorum*, ch. 6, p. 64.

Additionally, the golden milk is also good:

Take of: Non-sharp lye, waters of wild poppies,
of elderflowers, of each 3 ounces
vinegar of elder, oil of elder, of each 1½ drachms
egg yolks, 2 in number
saffron, camphor, of each 1 drachm.

Oswald Grembs in *Arbor ruinosa*, book II, ch. 1, p. 206.

The same author describes another water of *saffron* for erysipelas, ibid., p. 207.

When the inflammation is excessive, Gregor Horst recommends applying the following:

Take of: Frankincense, myrrh, of each ½ ounce
camphor, 2 drachms; *saffron*, ½ ounce
white wine, vinegar, of each 1 pound.

 | DOI:10.1163/9789004435292_025

Boil in a covered vessel, then moisten some linen cloths by soaking them in the vessel and apply them.

Möbius, *Anatomia camphorae*, ch. 23, p. 54.

II Arthritis

Publius said that just as scurfy hands are a sad business, thus gouty feet provide a sad freedom from work, in fact:

> *Qui jacet invitus, durius ille jacet*
>
> ["He who lies at rest unwilling, lies at rest more stiffly"]
>
> Eobanus, book III, *Heroidum*.

This is usually the common complaint of those suffering from arthritis.

What is arthritis?

Arthritis is the pain, with or without swelling, afflicting the parts constituting the joints. It is caused by mucilaginous tartar impregnated with salino-nitrous particles, or, according to others, by the inflammatory sudden flux of peculiar matter or by fluid which is sometimes rather thick, sometimes rather thin.

To calm the pain:

Take of: Flowers of ground-pine, rosemary,
mullein, violets,
red roses, of each 1 pugil
Austrian saffron, ½ drachm

gum ladanum, elemi, tacamahac, of each 1 drachm; aloes-wood, ½ drachm; root of orris, 2 scruples; mastic, frankincense, of each 2 drachms; calamite storax, 1 drachm.

Cut roughly.

Let it be marked: **Glieder-Rauch** ["Smoke for the limbs"].

For an anodyne:

Take of: Meals of bitter-vetch, peas,
lentils, of each ½ ounce
of linseed, 6 drachms

powder of flowers of chamomile, 3 drachms; yellow melilot, 2 drachms; *best saffron*, 1 scruple. Boil in water of chamomile-flowers, as much as suffices, until it reaches the consistency of pottage. When it has cooled down, add: egg yolks, 2 in number; oil of chamomile, 2 drachms. Make a cataplasm, apply lukewarm and replace when it dries. Scholz, *Consilia*, XII, p. 46.

Or

Take of: Core of white bread, as much as you will
saffron, a little.

Boil in cow's milk, and apply. Rolfinck, *Methodus medicinae specialis*, p. 863.

Or

Take of: Breadcrumbs macerated in milk, 1 pound
saffron, 2 scruples
flowers of chamomile, melilot, of each ½ pugil
seed-meals of linseed, fenugreek, of each ½ ounce
oil of lilies, 2 ounces.

Make a cataplasm. Gaspar Caldera de Heredia in *Tribunal medicum*, p. 228.

Take of: New milk, 10 drachms
core of a wheaten loaf, ½ drachm; barley-meal, 1½ ounces; hermodactyls, 6 drachms; *saffron*, 1½ drachms; oil of chamomile, 2 ounces. Oswald Grembs, *Arbor ruinosa*, p. 182.

Take of: Pulp of cassia, newly extracted, 1 ounce
oil of roses, 3 ounces
seed-mucilage of psyllium, 2 ounces
powder of hermodactyls, ½ ounce; *saffron*, ½ drachm. When the inflammation has subsided, add: oil of chamomile, powdered flowers of chamomile, or also meal of fenugreek.

Liberato Liberati in *Podagra politica*, first advice on arthritis, p. 266.

Take of: Dung of black cows, 1 pound
honey, 3 ounces
alum, 2 drachms
whole egg, *saffron*, of each ½ drachm
bole armeniac, 1 ounce.

With water of speedwell, of roses and vinegar, as much as suffices, make a cataplasm. At the end add: camphor, 1 drachm. To stop the fluxes. Finck in *Encheiridion dogmatico-hermeticum*, p. 184.

Sometimes, soft buckskin, with goat dung and *saffron*, placed on the gout takes away and soothes the pain. Brunfels.

When the pain increases:

Take of: Meals of barley, wheat, of each 1½ ounces

leaves of henbane, mallow, violets, of each 1 handful; flowers of chamomile, 1 handful; mullein, elder, of each ½ handful; rind of white poppy, ½ ounce; oil of earthworms, poppies, of each 1 ounce; opium, *saffron*, of each 1½ drachms.

Make a cataplasm, to be applied warm and to be replaced very often. Balthasar Timaeus von Güldenklee, *Epistolae medicinales*, p. 344.

Take of: Pulp of cassia, freshly-extracted, ½ ounce
seed-mucilage of psyllium, ½ ounce
saffron, 1 drachm
bean-meal, 2 drachms
opium, 1 scruple
oil of frogs, ½ ounce

Make a cataplasm, to be applied warm. It helps immediately. Thomas Bartholin, *Cista medica*, p. 423, from the rare *Curationes* of Tancred Lejel.

Or

Take of: Flowers of white mullein, chamomile, lilies,
henbane with heads, of each 1 pugil

green frogs, 10 in number; seed of white poppy [no quantity given]; river crabs, shattered with their shells, 20 in number; snails, earthworms washed in white wine, of each 4 ounces; badgers' fat, 6 drachms; oil of olives, 6 pounds. Boil everything in a copper vessel on the fire for two hours, then press vigorously. To the oil separated through boiling from the aquosity add:

Take of: *saffron*, 2 ounces
camphor, ½ ounce.

Pour everything in a glass vessel, and digest in water-bath for several days, or in the sun. Rodrigo da Fonseca, *Consultationes medicae*, pp. 350 and 351.

He who wishes to have an ointment:

Take of: Seed-mucilage of psyllium with scabious-water, ½ ounce
extract of saffron, 1 scruple

Mix. Oswald Grembs in *Arbor ruinosa*, p. 182.

Aëtius, *Sermones*, 12, ch. 36, describes this extraordinary medicament for soothing the pain, softening and removing the tophi:

Take of: Galbanum, frankincense, of each 2 drachms
vinegar, 6 drachms; *saffron*, 1 drachm
gum ammoniac, 3 drachms
stone pitch, 5 ounces
oil of roses, 1 ounce; wax, 6 drachms.

After melting the wax, pitch and oil of roses over a slow fire, strain them. Add galbanum, *saffron*, gum ammoniac, and frankincense to the melted ingredients removed from the fire. Teuber, *De arthritis*, section 8, ch. 6, p. 116, describes this extraordinary medicament as well.

In cases of extremely violent pain, juice of poppy with *saffron* and sheep's milk is most excellent. One has tried this: that a herring with *saffron* should be applied to the affected part of a person suffering from gout, the very salty skin of the herring having been removed, and the person will be healed. Then the vein should be cut, and the condition should disappear.

In cases of arthritis vaga, Friedrich Hoffmann in his scholarly work, *Methodus medendi*, p. 308, applies an ointment of Venice soap, spirit of earthworms, camphor and *saffron*, as much as suffices.

III Tumour

A frog, the subject of a fable and in love with the grand countenance of an ox, soon inflated by the wind of flattery, desired in every way to be like him. But the wretch, with this longing, created a bean inside her and as it swelled beyond the strength of the frog's little body and beyond its natural condition, the frog drew its last breath and burst dead.[1] In fact, every excess is an enemy of nature.

A tumour is an enemy of nature, and it is defined as the raising of a part, caused by bilious, pituitous, melancholic blood or by superfluous nutritious fluid. What is a tumour?

When it tends to suppuration, to take away the pain, apply *powder of saffron* to the area below the tumour and the following plaster on top of it:

Take of: Egg yolks, 3 in number
complete oil of roses,
butter, of each as much as suffices
saffron, 1 scruple
yeast, as much as suffices.

Mix everything to form a plaster. Rodrigo da Fonseca, *Consultationes medicae*, p. 127.

1 This story is based on one of Aesop's Fables, number 376 in the Perry Index.

To remove a tumour in the testicles, and soothe the pain, François Joël applies this cataplasm:
Take of: Pulp of mallow, pellitory of the wall, of each 1 ounce
wheat-bran,
mill-dust collected in the walls of mills, of each ½ ounce; powder of fenugreek, dill, of each 3 drachms; *saffron*, 1 drachm. Boil in wine to the consistency of pottage, adding oil of chamomile, as much as suffices. Make a cataplasm.

Against the same tumour, the following ointment is effective:
Take of: Jew's-ear fungus of elder, 20 in number
Infuse one night in water of the same, 18 ounces; boil to expression, add:
seed-mucilage of quince, 2 ounces
oils of elder, roses, of each 1½ ounces
[oil of] *saffron by distillation*, ½ scruple.
Mix. Make an ointment.

IV Scrofula

In the same way as Pyrrhus, King of Epirus, once cured those suffering from a bad spleen only with the touch of the big toe of his right foot, see Fregoso, book I, ch. 6, from Plutarch's *Pyrrhus*, I, thus, nowadays, the Kings of France, thanks to a singular quality granted to them by God, cure the struma only with a touch and, by making the sign of the cross above it, they pronounce these words:

Rex Te tangit
DEUS *Te sanet*

["The king touches you,
GOD heals you"]

What is scrofula?

Strumas are swellings or whitish flesh, easily growing, enclosed in a person's own flesh, at times hanging and not firm, at other times immovable and adjoining the trachea. They are caused by excessive nutritious fluid in the glands.

The following powder helps quickly with this condition:
Take of: Sea-sponge, burned by the sun, 3 ounces
palla marina, 2 ounces

eastern saffron,
pellitory of Spain, salt, of each ½ ounce

ginger, 1 ounce; galangal, parsnip, of each ½ ounce; long pepper, frankincense, cubebs, of each 2 drachms; cuttlefish, pumice, of each 1 ounce, cypress-nuts, 2 drachms. Pound everything very finely and let the patient take it whenever he wishes to. Stocker, *Empyrica*, book I, ch. 23, p. 58.

Or

Take of: Powder of sponge, dried, and not burned,
in an oven after the bread has been taken out,
the vessel having been carefully
luted, 3 drachms
Mynsicht's arcanum duplicatum,
powder of scorzonera, of each 1 drachm
magistery of spine of vipers,
saffron, of each ½ drachm.

Powder for strumas

Mix. Make a powder. Dose: 1 scruple to ½ drachm when the moon is waning.

Additionally, Mynsicht's *plaster of galbanum and saffron* is also effective.

V Breast Inflammation

Often great accidents happen in short intervals of time; just as nothing should be feared, thus nothing should be disregarded. Breast inflammation is per se less dangerous with respect to the other tumours, it can, however, become dangerous on account of the parts affected, since the same often degenerates into schirrus or cancer.

What is breast inflammation?

Breast inflammation is a hard, hot, painful red swelling with fever, caused by blood flowing more abundantly because of the retention of the lochia, or stagnating there because of the abundance of milk.

In order to prevent breast inflammation, and the coagulation of milk (vor das ungesegnet ["against the unseen"] as they call it), women take a rough tissue, which has not been exposed to the sun, and immerse it in water in which *saffron* and camphor have been dissolved. They dry it and apply it dry to the breasts.

If the inflammation and the coagulation of milk are in progress, use this:

Take of: Blue paper anointed with fresh butter, sprinkling over it shavings of liquorice with *whole saffron-flowers*, and thus apply to the breasts.

In cases of breast inflammation, François Joël prescribes this cataplasm to dissipate and soften:

Take of: Core of a loaf of bread made of the finest flour, 2 drachms.
Boil in fresh milk to make a cataplasm, add:
saffron, 1 drachm; myrrh, 2 drachms
egg yolks, 2 in number.
Mix. Make a cataplasm; book III, section 3, p. 298.

Philip Grüling uses this to cause the swelling to ripen:

Take of: *Saffron,* 1 drachm
egg white and 1 spoonful of March-, April-, or May-butter, mix together, spread over leather or over a hempen cloth and apply to the breasts.[2]

For a greater suppuration:

Take of: Mallow, 1 handful
violets, ½ handful
figs, 5 in number.
Boil until soft, pound, add:
powder of root of marshmallow,
seed-meals of linseed, fenugreek, of each 1 ounce
Add:
egg yolks, 3 in number
saffron, 1 scruple.
Mix. Make a cataplasm. Sennert.

VI Burns

When the sun is getting close to sunset, shadows increase to such an extent that they turn into night with the end of the day.

Often the shadows of an imminent gangrene, when natural heat declines, follow ill-treated burns, to the ruin of the whole body.

What are burns?

Burns are the combustion of body-parts caused by fire, boiling hot water, metal or gunpowder, appearing with or without pustules.

2 March-, April-, or May-butter: butter made in that particular month from milk produced by cows grazing the particular grasses and flowers growing during that time, as opposed to milk from hay-fed cows in winter.

Balthasar Timaeus von Güldenklee, *Casus medicinales*, book VI, p. 293, applies this ointment to draw out the empyreuma and, at the same time, to soften the dry skin:
Take of: Bleaks freshly caught in running water, 1 pound
Fry in a frying-pan with May-butter, as much as suffices. Pour the fat over extremely cold water, separate it when it has coagulated. To it add: basilicon ointment, 1 ounce; oils of frankincense, roses, sweet almonds, of each ½ ounce; egg yolks, 2 in number; *saffron*, 1 scruple. Mix. Make an ointment.

Or
Take of: 6 egg yolks
oil of roses, 2 ounces
saffron and bole, of each a little.
Mix. Make an ointment.

VII Gangrene

When public disasters are not perceived on a large scale, it is certain that extreme destruction is looming for entire states.

In a sick body, doctors consider it to be a deadly sign when the diseased himself does not feel the disease and, if asked, is not able to tell what part is in pain. This happens with gangrene, as it turns into sphacelus.

Gangrene is decomposition beginning with the lividity of the part affected, then blackness and numbness. It is caused by an excessive flow of humidity and excessive refrigeration, external heat, a strong ligature impeding blood circulation, a malignity, etc. What is gangrene?

In this instance, Ruland in *Pharmacopoeia nova*, tech. 3, p. 56, has experimented with the following:
Take of: Cow dung, 1 pound
oil of roses, 4 ounces; vinegar, 3 ounces
saffron, ½ scruple.
Mix for a plaster. Apply warm every three hours. It releases the heat, soothes the pain and removes the swelling.

VIII Ecchymosis

A thief recognizes a thief, a wolf recognizes a wolf.

Not always does brother recognize brother, especially when the face of one of them has been somehow disfigured with bruises.

Something similar happened to that Olympic wrestler who, because while wrestling, had been maimed in the face so much that he simply looked different from himself, was deprived of his father's inheritance. Since a portrait of him was produced to the judge some time before, his brother demonstrated that he was not his brother. *Theatrum vitae humanae*, volume XVI, p. 2496.

What is a contusion?

A contusion is the profusion of blood condensing under the skin happening with an attack without a wound, because of something external.

The following cataplasm is unparalleled in loosening and drying the matter:

Take of: Roots of both kinds of consound, of each ½ pound
chamomile, melilot, of each 2 ounces
saffron, ½ drachm; bean-meal, 4 ounces
unsalted butter, 5 ounces; fenugreek-meal, 4 ounces

Boil both roots until they are moderately cooked, then pour in the remaining ingredients, after adding juice of wormwood and cumin, of each 1 ounce. Epifanio, *Empyrica*, p. 411.

Mattioli shows that, in cases of falls from height, especially in infants, nothing is more effective than *saffron* put in vinegar, until it is tinged with it, and taken warm.

Or

Take of: Bole armeniac, terra sigillata,
madder, choice rhubarb, of each 1½ drachms
myrtillus, 1 drachm; *saffron*, costus, of each ½ scruple

Mix. Make a powder. Ferdinando Epifanio in *Empyrica*, p. 450.[3]

IX Wounds

Those whose wounds in the body heal easily are similar to the pike in England. The buyer agrees on a price for the pike, provided that it is fat, but the

3 Again, the author of *Empyrica* was Nicola Epifanio, not Ferdinando.

intestines are examined by making a wound. If he does not like it, the pike, stitched, often also not stitched, is returned to the water, and the wound, despite this, heals and the pike stays alive.

Some attribute the cause of this to fish-ponds in which tenches are also kept; they think that wounds coagulate and compose thanks to the rubbing against those, just like sarcocolla.[4] Cardano, *De subtilitate*, book X, and Della Scala, *Exercitatio* 201, ascribe this to the glutinous flesh of the pike, evidence proved by the fact that the flesh of the pike, among different foods, sticks when it comes in contact with the fingers.

A wound is a bloody dissolution of the whole, caused by an external cause, by means of a cut, a puncture, the bullet of a rifle, the bite of a rabid animal, etc. What is a wound?

Fabry von Hilden, *Observationes chirurgicae, centuria* 2, observation 79, p. 150, proposes this digestive for wounds:

Take of: Turpentine washed in scabious-water, 2 drachms
new wax, gum elemi, of each 1 drachm
oils of roses, sweet almond, of egg yolks, of each ½ ounce
saffron, 1 scruple; egg yolk, 1 in number.
Mix. Make an ointment.

Mesue's golden ointment, containing *saffron*, closes the wounds in a healthy and reliable way. See Pierre Potier, *Pharmacopoea spagyrica*, book III, section 9, p. 575.

When the wound is inflicted by a rabid dog:
Take of: Spanish flies, the legs and heads cut off, 1 drachm
cleaned lentils, ½ drachm
saffron, spike, cloves, of each ½ scruple.
Mix. Dose: ½ scruple with wine in the morning for 3 or 4 days, until the patient urinates blood, and he will be healed. Epifanio, *Empyrica*, p. 431.

When a puncture wound is on the nerves:
Take of: Flowers of St John's wort, 1 pound
extremely old oil, 1 pound.

4 An alternative name for the tench (*Tinca tinca* L.) is doctor fish. Walton, 1653, pp. 175–176, writes that the tench is, "observed to be a Physician to other fishes … it is said, that a Pike will neither devour or hurt him [a tench], because the Pike, being sick or hurt by any accident, is cured by touching the Tench, and the Tench does the like to other fishes, either by touching them, or by being in their company."

Mix. Expose for several days to the rays of the sun, then press vigorously. Then add the same quantity of flowers of St John's wort; turpentine, 2 ounces; granulous resin, euphorbium, castoreum, of each ½ ounce; *saffron*, 1 drachm; worms washed with white wine, 2 handfuls. Mix, distil according to Art and store. Instilled warm, it cures punctures of the nerves and of the nervous parts, it heals contusions of the joints, and flatuous and pituitous tumours; and this is the Grand Duke of Tuscany's oil against cramps. See Prévost, *Selectoria remedia*, p. 38.

x Ulcers

A dog bite is considered so dangerous that he who has been bitten once by a dog, for pestilential contagion, is known to spoil the broodings of hens, bearings of sheep and cause abortions with his intervention, and to make the wounds of another worse with his passage.

Giovanni Pierio Valeriano

Even more dangerous is the urine of a rabid dog, so that, if someone has thrown it on his own urine, he is made more sluggish at begetting by a numbness in the loins. Moreover, the same urine, being somehow trodden upon, is extremely harmful to those who have ulcers, thus, one must keep clear of it at all costs.

What is an ulcer?

An ulcer is a dissolution of the whole with pus, caused by corroding, pungent, sharp humours, or by a wound or an abscess.

Against tibial ulcers, Dr Philip Jakob Sachs von Löwenheim recommends tibial bands with water of *saffron* and elderflowers.

For cleansing the ulcers, Pierre Potier, *Observationum et curationum, centuria* 2, p. 132. commends the following:

Take of: Pine resin, ship pitch, of each 2 ounces
clear turpentine, 1 ounce
2 egg yolks
round birthwort, frankincense, of each 1 drachm
saffron, 1 scruple; oil of roses, a little.

Mix. Make an ointment.

When a serpiginous ulcer has turned into a fistula, Thomas Keßler in *Chymicus processus*, p. 153, commends incredibly his balsamic water, useful for pestilential buboes and carbuncles as well:

Take of: Leaves and flowers of rosemary,
bay, sanicle,

speedwell, tobacco, myrtillus, plantain, of each [no quantity given]. From these distil a water, and if you have 6 pounds, add: Venice turpentine, 1 pound; gums of ivy, cherry-tree, of each 3 ounces; frankincense, myrrh, true mummy, of each 2½ ounces; aloes, 3 ounces; cloves, mace, of each 1 ounce; sugar, ½ pound; *saffron*, ½ ounce. Mix all and distil with sand according to Art. Leave the distillate be until it cools down. Take 3 drops of the oily water of this and wash the fistula with them.

Or
Take of: Ordinary oil, 2 pounds
sal-ammoniac, 1 drachm; fossil pitch, 2 drachms; *saffron*, 1½ drachms; boil in a water-bath to the consumption of one third of it, and keep linen patches imbued with this oil on the fistula for four hours every day. Experimented by Jean Prévost, *Selectoria remedia*, p. 69.

Rodrigo da Fonseca, *Consultationes medicae*, p. 31, vol. 2, describes a medicament with *saffron* for cleansing a fistula and causing it to dry.

XI Synovia, in German das Gliedwasser ["The Limb Water"]

The most dangerous springs are those which entice the eyes with clear waters, but bring ruin. The springs to be feared less, instead, are those which show with the appearance of their waters that they should be avoided. Pliny, book XXXII, ch. 2.

With synovia, there is a discharge of transparent liquid. But who will not fear this in a wound more than any secretion of pus, when any consolidation of wounds is in vain because of its discharge?

For internal use, Dr Lorenz Scholz, in his work, commends a magistery of crab's-eyes or a little potion from water of nettle, burdock, into which cast in crab's-eyes, frankincense, egg shells, of each 2 drachms; *saffron*, 1 drachm. Take for 5 days continuously, see *Gammarologia curiosa* by Dr Philip Jakob Sachs von Löwenheim, p. 830.

CHAPTER XXIII

The Mechanical Uses of *Saffron*

The Orator of Chio is a proof that the red dye for the hair (for the dyeing of which *saffron* is of no little importance) was hated exceedingly by the Spartans. When old age made him ashamed, he had dyed his grey hair with potions.

But Archidamus, becoming Prince of Sparta, disconcerted him publicly, saying, "For heaven's sake, what soundness of speech is this man going to have, who carries around falsehood not only in his soul but also on his head."

Alessio Piemontese, book IV, *De secretis*, p. 204, describes an *oil of saffron* of this kind for dyeing the hair blonde.

The following is similar:

Take of: Middle bark of elder,
flowers of broom, *saffron*,
egg yolks, of each as much as suffices.

Boil in water and collect the supernatant, and then anoint the hair.

Or

Take of: Barley-straw, liquorice-rind,
shavings of box-wood, leaves of box-tree,
saffron, cumin, of each as much as suffices
lye of brushwood ashes, as much as suffices.

Boil everything well, and make lye. If the hair is washed very often with this, it becomes very blonde. It imitates the colour of gold. Ibid, pp. 178 and 179.

A golden solution, without gold, for writing can be thus made with *saffron*:

Take of: *Juice of fresh saffron-flowers* or, if you do not have that:
crumbled saffron, best and bright orpiment, of each one part.

Mix and bruise with goat gall or wolf urine, which is better. Then, after it has been placed in a vessel for some time, take out and store the solution for writing and painting. Wecker, *De secretis*, book V, p. 617; Piemontese, *De secretis*, book V, p. 253.

For another solution with *saffron* for gilding hides, with which sometimes other things, sometimes carpets and tapestries are made, see Piemontese, *De secretis*, book XVI, p. 627.

 | DOI:10.1163/9789004435292_026

To tinge sugar with a bright colour:
Take of: *Saffron*, 1 drachm
Pour in 1 pound spring water, heat it for a day and a night, then strain and with this water boil the sugar. Alessio Piemontese, *De secretis*, p. 166.

Questions Concerning *Saffron*

QUESTION I.
Is *saffron* a remedy against monkshood?[1]

Nobody who has pondered what we have said in the chapter *On the virtues of Saffron* & § *Plague* will doubt that *saffron* has a great power against poisons.

Pieter van Foreest, book XXX, *De venenis*, in Scholz, p. 122, agrees with us as he lists *saffron* among other ingredients overcoming the poison of monkshood.

A letter from the excellent Dr Michael Ettmüller, MD, practitioner and most esteemed colleague in Leipzig, seems to confirm this even more. In this letter, recently given to us, he reports that he heard from one of his friends, a doctor in Poland, that the inhabitants of that State, with great pleasure and harmlessly, add to salads flowers of monkshood, which elsewhere are said to be extremely poisonous.[2] Upon hearing this, I was stupefied and I thought that maybe they were different flowers, in particular those of the larkspur, since there is a great similarity between the two and since the flowers of

1 Monkshood, *Aconitum napellus* L. (Ranunculaceae) is a notoriously poisonous plant, without a specific antidote, and it still occasionally causes fatalities. It originates from alpine areas in Europe, and is a valued garden ornamental for its showy blue flowers. All parts of the plant (especially the root) are toxic, containing alkaloids including aconitine. Accidental cases of aconitine poisoning from monkshood have occurred e.g. after foraging wild plants and mistakenly eating monkshood (Pullela et al., 2008). Symptoms can begin minutes after exposure and in severe cases, death results from ventricular arrhythmia (abnormal heartbeat rhythms originating in the ventricles of the heart). Treatment therefore focuses on management of arrhythmia with anti-arrhythmia drugs e.g. lidocaine (Furbee, 2009). Historical treatments for monkshood poisoning included the use of *Aconitum anthora* L. "mithridate monkshood" or "wholesome monkshood" (Gerarde, 1597, book II, p. 819), also known as "counterpoison monkshood" (Parkinson, 1629, p. 215). Parkinson (*idem.*) related how monkshood flowers were appreciated for their beauty, and were "laid among greene herbes in windowes and roomes for the Summertime" and he goes on to caution, "beware they come not neare your tongue or lippes, lest they tell you to your cost, they are not so good as they seeme to be".

2 There could be geographic differences in the levels of aconitine in different monkshood accessions, either because of genetic or environmental factors, or a combination of both. Ingolfsdottir & Olafsson (1997) investigated how, despite the toxicity of monkshood, Icelandic children were able to eat the flowers' nectarines as a snack, with no apparent ill effects. It was discovered that the nectarines and whole flowers of the particular *A. napellus* ssp. *vulgare* accession tested had only a very low aconitine content.

 | DOI:10.1163/9789004435292_027

monkshood do not grow so much in gardens as in the mountains.[3] But, writing again, the aforementioned Doctor confirmed the previous things, and he ascribed the reason for the harmless assumption to the extraordinary power of *saffron*, which corrects the virulence of the monkshood, since Poles like *saffron* very much and often add 1 ounce of it in broths in one single time and season fish with it.[4]

Unless someone maybe wants to say that the flowers of monkshood in Poland are not as deadly as those in Germany or Italy? For example the Polish poppy seed, assumed up to 1 ounce and more does not harm at all; in contrast, it has been observed by many that in the same dosage in Italy, it hurts greatly.[5]

QUESTION II.
Does *saffron* prevent intoxication or, rather, cause it?

Dioscorides provides a reason for doubt as he wrote that *saffron* prevents intoxication if drunk with raisin-wine. And Pliny does not depart from this path; in fact, not only did he believe that *saffron* fought drunkenness if drunk, but he also wrote that *crowns of saffron* alleviate drunkenness.

Others, however, like Serapion, following Rhazes, maintain that *saffron* intoxicates if mixed with wine. Johann Michael, a very famous practitioner from Leipzig, agrees with him. He demonstrated that wine in which a tiny bit of *saffron* has been mixed, intoxicates quickly. *Saffron*, in fact, cheers the heart of man, like wine. It vivifies the spirit, it enlivens all senses, therefore it is not surprising that a wine of such kind, if drunk, makes a man drunk more quickly.[6]

---

3 Larkspur, *Consolida regalis* Gray (syn. *Delphinium consolida* L.) (Ranunculaceae) is superficially similar to monkshood.

4 Hill, 1751, p. 437, also quotes Ettmüller ("Etmuller") and proposed two ideas why 1 ounce doses of saffron in Poland (cf. the contemporary normal "medical" dose in the UK of 5–10 grains) could be tolerated: (a) Polish saffron was not as strong as English saffron; (b) the Polish population had become tolerant to large doses "that would be fatal to others unaccustomed to them" by long-term usage. However, it does seem highly unlikely that this quantity of saffron was used to make a single meal: the flavour would be highly unpalatable and the cost unimaginably high.

5 Poppy seed remains an ingredient of traditional Polish cookery, e.g. in Polish poppy-seed cake (Strybel & Strybel, 2005).

6 Harrison, 1587, p. 393, tells how high doses of saffron in food could also cause intoxication: "I have knowne some, that by eating onelie [only] of bread more than of custome streined [stained] with saffron, have become like droonken [drunken] men, & yet otherwise well known but to be competent drinkers."

If, however, it [saffron] prevents intoxication, one must believe that perhaps those spirits that hinder the spirits of wine are made more abundant, so that the latter cannot act. In fact, the spirits of *saffron* are more alike our spirits than the spirits of wine; the spirits of *saffron* are, indeed, extremely favourable to the heart.

Or, because *saffron* is beneficial to the stomach, on this account, maybe it refrains the excessive rising of vapours to the head. In fact, those having a weak brain and a burning stomach are intoxicated more easily. Jason Pratensis, *De cerebri morbis*, ch. 20, p. 320.

Or, surely, *saffron* prevents intoxication because it enlarges the vessels, heart and veins, since those having a hot and thicker body constitution and narrower veins are intoxicated more quickly. Houllier, *Commentarii Hippocratis*, section 5, § 5.

If *saffron*, however, exceeds our spirits in quantity (as said very often), by dissipating them, it kills with laughing.

QUESTION III.
Does *saffron* induce sleep or wakefulness?

Pliny writes that *saffron* induces sleep, and so does Mattioli in *Herbarium*, p. 18.

Some, who maintain that a wreath of *saffron-flowers* placed around the head induces sleep, confirm this.

Others, instead, believe that it fights sleep-inducing conditions, among them Isaac Israeli, *De dieta et herbis particularis*, p. 481; Bock; Freitag, *Aurora medicorum*, p. 502 et seq., who recommends *saffron* with sleep-inducing conditions, especially with lethargy.

For a reconciliation of both sides, we agree with the first, that *saffron* is anodyne in the highest degree, as said in Simon Paulli, *Quadripartitum botanicum*, p. 52, therefore it is able to induce sleep.

It should be added that *saffron* is methodically mixed with other soporific and narcotic medicaments, so that, because of its penetrative virtue, it causes the narcotic medicaments to penetrate more and to work more quickly.

The second side of the argument does not present any difficulty: *saffron* is, in fact, hot and dry, therefore, it is found highly advantageous in cold and moist conditions such as sleep-inducing conditions.

Secondly, *saffron* causes digestion, hence, it will be most serviceable in digesting and separating phlegm.

Thirdly, it enlivens the senses, which, benumbed, became inactive in these conditions.

QUESTION IV.
Does *saffron* tinge a foetus in the womb?

The Philosopher, in the *Liber predicamentorum*, says that it is not useless to doubt about individual things, so that what is true, sound and genuine indisputably stands out from one side of the question.

Amato Lusitano seems to say something great and worthy of admiration, as in *Curationum medicinalium centuria*, 5, p. 71, he reports that a woman, from drinking a medicament containing *saffron*, gave birth to two little girls of a yellow colour on the outside. He ascribes the reason for this pigmentation to *saffron* itself, proof having been found in the fact that the colour disappeared with a bath in water.[7]

A cause for doubt was not absent from us, since all eatables, either red or yellow or of whatever colour they are, at the end, when their original and very own colour has disappeared, become white.

Furthermore, since, as Marcus Tullius says, nothing is sweeter than knowing everything, this year, and specifically in Lent, I subjected Amato's claim to the anvil of truth.

I took a female dog, three or four weeks pregnant, and during the remaining time of the gestation I gave her food and drink tinged with *saffron*, until maybe eight days before the delivery. In the last three days, I also added up to 2 and 3 drachms of *saffron* to broths.

Even though initially the dog rejected these *saffron*-flavoured meals, then, however, hunger becoming more pressing, it ate them rather quickly.

At the time, I was fitly engaged in the public dissection of the body of a woman executed by beheading; thus, at the end, having successfully carried on the preparation and exposition of all the parts, I devoted the aforementioned dog to public dissection too, so that the truth of what was said by Amato, and by some doctors agreeing with him, could be brought to light.

I call upon the witnesses present at the time, the illustrious Dr Philip Christoph von Lichtenstein, Count of the Holy Roman Empire, etc., and the illustrious Dr Johannes Antonius, son of the Baron of Questenberg, etc., the illustrious Senate of Brno, and the other doctor observers assisting.

7 The belief that long-term use of saffron could colour a foetus yellow was maintained by writers long after Hertodt's time, e.g. Pereira, 1840, p. 676. A case of "chrysanthropy" (the birth of a golden yellow baby due to its mother's saffron consumption) was reported in the 1906 *Revue de pharmacologie médicinale* (cited in Thiercelin, 2003, p. 18).

Thus, having cut the entire region of the abdomen of the dog, as well as the uterus itself up to the placentas and having dissected those as well, I took the colliquament, which seemed tinged of a not inconsiderable *saffron* colour, and I placed it in a bowl and I instructed to bring it near to hardening with [the heat of a] fire. In the meantime, attention was to be paid to the puppies, rolling themselves here and there with a pleasant appearance: they were of a variety of colours from black to white and the whiter parts of their skin were completely tinged in a yellow colour.

Therefore, it appeared more clearly in the light of day that what Amato Lusitano said was true, that *saffron* tinged the foetus inside the uterus too, and that the same was employed with excellent reason, like a king, for driving the foetus and the menstruation out.

But it was not apparent how the chyle was white and not tinged by *saffron* in the lacteal vessels scattered through the mesentery. However, how it has been granted colouration, and not deprived of it, is an entirely different investigation.

Unless we want to say that extremely thin particles of *saffron*, in which the pigmentation is rather consistent, after the thicker chylous residues have been left behind, and prevailing over them, with a quick contraction of the heart, proceed furiously towards the heart through the thoracic duct, when the heart is very much pleased by them. But because the same colouration is not visible in the blood, one must believe that, because of the rich and intense redness of pigmentation of the blood, it cannot be perceived there, as the colouration deepens it. In fact, *saffron* possesses a blood-coloured pigmentation and abounds greatly in fixed sulphur and, therefore, because of the very compressed and compacted pores of the same, it hardly undergoes a loosening of its pigmentation, whilst that which is fumigated with sulphur looses colour. Maybe because *saffron* is a solar plant, it should add colour to sulphur, the sun of metals, when, instead, the other plants having very wide pores easily change on the surface, also from the solar rays, their colouration having disappeared.

Alternatively, if we want to agree with the Chemists, the colouration of *saffron* is directed only into the chyle, not otherwise. It usually comes into very red corals, when they are touched by the flame superficially on the thin shells and they repel it and become white, but by the succeeding degrees of the fire they are changed into reddish yellow to yellow to intense redness (which initially they had hidden), whence, only then, the colouration becomes extremely red. But since the colouration of *saffron* shows itself to the sight, again one must infer that this happened in the colliquament. When blood impregnated with *saffron* is carried through the arteries to the placentae, and there, purified by the glands, clearer colliquament comes out, and thus, just as in the arteries, it cannot disguise and hide the colouration of *saffron*, but it is forced to show it.

If someone is of the opinion that the chyle is carried through the lacteal vessels to the womb, on this account we disagree with him because no vessels directed there have been found. And the lymph vessels cannot oversee that function, considering their use.

But! While we are mentioning the lymph vessels, I would like to add as a conclusion an observation of the lymph vessels that I made in presence of very distinguished men, Drs Scholz, Sattler, Renck and others. Some anatomists of great importance deny that the lymph vessels surround the brain and, through the membranes, spread to the brain itself.

Observation on the lymph vessels

By experience I demonstrate the opposite. In fact, some years ago in Padua, the body of a child who died from hydrocephalus was given to me from the Casa di Dio by the illustrious Count Geronimo Frigimelica, most magnificent and deserving protector of the German nation. Having opened the child's cranium, I found lymph vessels inside and outside the brain, crawling here and there like the calamus scriptorius, some broken, some still full of lymph. And no-one can argue that these were maybe arteries and veins, as, because of the abundance of serous humours, with which this subject very much abounded, I was able to distinguish lymph vessels from arteries and veins very well.

And thus, for the greatest glory of GOD,
The prosperity of fellow-men, and the growth
Of the Republic of Letters of the *Curiosi*,
These things be said about *saffron*.

Illustration of Different Kinds of Crocuses, and Explicative Table

FIGURE 3 Different kinds of crocuses, see Explicative Table below for more information. This image is from a copy of *Crocologia* in the Wellcome Collection London. IMAGE PUBLISHED WITH PERMISSION OF PROQUEST. FURTHER REPRODUCTION IS PROHIBITED WITHOUT PERMISSION.

Explicative table

A. Ordinary green saffron crocus.[1]

B. Ordinary flowering saffron crocus exhibiting an autumnal flower, to which the real crocus with white flower is associated.

1 Green meaning here the same as the horticultural term "in the green", i.e. in leaf, without flowers.

 | DOI:10.1163/9789004435292_028

C. Autumnal wild saffron crocus with purple flower, sub-classified into saffron crocus with violet flower, Pyrenaeus with dark purple flower, biflorus, montanus.

D. Broad-leaved saffron crocus with completely white flowers. Sub-classified into broad-leaved saffron crocus with striped white flowers, broad-leaved saffron crocus with an extremely large purple flower and a smaller one, with multicoloured white flower, with multicoloured purple flower, with pale flower, with golden flower.

E. Broad-leaved saffron crocus with a big flower with purple stripes.

F. Multi-flowered vernal saffron crocus with purple-blue flowers. This includes also the multi-flowered vernal saffron crocus with several white flowers arranged in groups.

G. Multi-flowered vernal saffron crocus with very large leaves lying low from the ash-coloured bulb, like a sheath on the waxy tawny colour of the flowers, rising up in groups with the corolla totally upright.

H. Narrow-leaved vernal saffron crocus with a big white flower, among which narrow-leaved vernal saffron crocus with variegated white flower, with purple flower, with variegated purple flower, with reddish-yellow [orange] flower, Spanish with violet flower, Spanish with dark purple flower.

APPENDIX 1

Hertodt's References

Crocologia documents and authors cited by Hertodt. No editions are listed after 1671 as Hertodt could not have accessed them for *Crocologia*. References in speech marks are those for which Hertodt's given title does not match books in the WorldCat or CERL databases.

Acts of the Apostles. [A book of the Bible].

Actuarius, Johannes Zacharius. Byzantine physician; lived 1275–c. 1328.

Ader, Gulielmus. Author of *Enarrationes de aegrotis, et morbis in Euangelio*; died in 1638.

Aëtius. *"Sermones." De placitis philosophorum*.

Albert the Great [Albertus Magnus]. *De animalibus*.

Alciato. *Emblemata*.

Alexander of Tralles [Alexanderus Trallianus]. "Book I". Physician and author of *Twelve books on medicine*.

Altensteig, Johann (listed incorrectly as Altenstaig in Hertodt's original text). Writer of dictionaries; lived 1480–1523 (CERL).

Amato Lusitano [Amatus Lusitanus]. *Curationum medicinalium*. Editions published 1560–1570, Lyon.

Angelus, Daniel. *Catalogus medicamentorum spagyricorum pharmacopoeia spagyricae ... appendix de compositione medicamentorum ...* 1667, Gdansk.

Archidamus. Probably the king of Sparta who reigned from approximately 476 BC to 427.

Aristotle. *Categories. Liber predicamentorum*.

Aristotle. *De historia animalis*.

Aristotle. *De mirabilis ausculationibus*.

Aristotle. *Metaphysics*.

Ausonius. Fourth century Roman poet.

Avicenna [Ibn Sina]. Persian polymath; lived c. 980–1037.

Bacon, Francis, Viscount of St Albans. *Historia vitae et mortis*. Editions published 1623–1663. 1st ed. published in London; later editions also in mainland Europe.

Banzer, Marcus. Dean of the Faculty of Medicine in Wittenberg. Quoted by Sachs von Löwenheim in his *Gammarologia* (see below).

Barbette, Paul. *Praxim Barbettianam cum notis et observationibus Frederici Dekkers*. 1669, Leiden.

Bartholin, Thomas [Bartholinus]. *Cista medica Hafniensis*. 1662, Copenhagen.

Bartholomaeus Anglicus [Bartholomew the Englishman]. *De proprietatibus rerum*. In 19 books.

 | DOI:10.1163/9789004435292_029

Bauhin, Caspar/Gaspard [Bauhinus, Casparus]. Swiss botanist; lived 1560–1624 (CERL).

Bauhin, Johann. *Historia plantarum universalis.* Two volumes. 1650 & 1651.

Behr. A doctor of Strasbourg (Hertodt, 1671).

Besler, Basilius. *Hortus Eystettensis.* 1613, Nuremburg.

Besold, Philip. Possibly an error and should be Hieronymus Besold, 16th century German theologian and writer of commentaries.

Bils, Louis de [Bilsius]. Dutch anatomist; lived 1624–1670 (CERL).

Bock, Hieronymus [Tragus]. German botanist and physician; lived 1498–1554 (CERL).

Böckel, Johann. *"Regimen pestis"* possibly part of *De peste.* 1577, Heinrichstadt.

Bodin, Jean [Bodinus]. *De magorum daemonomia.* 1591, Strasbourg.

Boë (Sylvius), Franz de le [Sylvius, Franciscus]. Dutch physician & scientist; lived 1614–1672. (CERL).

Bonacciuoli, Luigi [Bonaciolus, Ludovicus]. *De foetus formatione.* 1650, Leiden.

Book of Job. [A book of the Bible].

Book of Kings. [A book of the Bible].

Borel, Pierre [Borellus]. *"Observationes"* possibly *Observationum microscopicarum centuria.* 1656, The Hague.

Bottoni, Albertino. Physician in Padua (Hertodt, 1671); died c. 1596 (CERL). Prolific author.

Brodaeus, Johannes. *Miscellanea* book VI. 1555, Basel.

Brunfels, [Brunfelius]. Otto Brunfels, German botanist and theologian; lived 1488–1534 (CERL). Prolific author.

Brunner, Balthasar [Brunnerus]. *Consilia medica.* 1617, Halle an der Saale.

Caesalpino, Andrea [Caesalpinus]. "Book I." Italian botanist and physician, director of the Botanical Gardens at Pisa; lived 1524/5–1603 (CERL).

Caldera de Heredia, Gaspar [Calderae, Casparis]. *Tribunal, medicum, magicum, et politicum.* 1658, Leiden.

Camerarius, Joachim the Younger. *Hortus medicus et philosophicus.* 1588, Frankfurt.

Camerarius. *"Epitome"* possibly Mattholi & Camerarius, Joachim the Younger, *De plantis epitome utilissima.* 1598, Frankfurt.

Camerarius. *"Memorabilium"* possibly Camerarius, Johann Rudolf, *Sylloges memorabilium medicinae et mirabilium naturae arcanorum.* Editions published 1614–1652.

Capelluti, Rolando [Capellutius, Rolandus]. *"Tractatus chirurgicus de peste"* possibly *Tractatus de curationes pestiferorum apostematum.* Editions published 1468–1648.

Capivaccio, Girolamo [Capivaccius, Hieronymus]. Professor of Medicine at Padua; lived 1500–1589 (CERL). Quoted the letters collected by Scholz, see below.

Cardano, Girolamo [Cardanus]. *"Consilium de respirationis difficultate"* possibly *Hieronymi Cardani Quaedam opuscula ... Consilium pro illustri Hieronymo Palavicino Principe curtis majoris, difficultate spirandi laborante ...* 1559, Basel.

Cardano, Girolamo [Cardanus]. *De subtilitate.* Editions published 1550–1664.

Chrysostom, St John. *"De mente irata et tranquilla."*

Cicero. *In Verrem.*

Cicero. *Oratio ad Quirites.*

Cicero. *Somnium Scipionis.*

Cicero. *Tusculanae quaestiones* II.

Clusius [L'Écluse, Charles de]. "Book II" & "Book V." French botanist, Professor of Botany at Leiden; lived 1526–1609 (CERL).

Coler, Johann [Colerus]. *"Gartenbau"* possibly *Oeconomia ruralis et domestica ... Algemeines haussbuch und kurtze beschreibung von haußhalten, wein-, acker-, garten-, blumen- und feldbau.* Editions published 1645–1665.

Costeo, Giovanni [Costaeus]. Possibly Giovanni Costeo, Italian Professor of Medicine and Philosophy; lived 1500–1603: or Giovanni Francesco Costeo, Italian physician; lived 1565–1608. (CERL). Medical books by Johannes Costaeus published 1564–1658 (WorldCat).

Crato, Johannes [Krafft, Johannes]. Chief Physician to three Holy Roman Emperors (Hertodt, 1671).

Crescenzi, Pietro de' [Crescentium]. *De agricultura.* Editions published 1478–1605.

Croll, Otto [Crollius]. *Basilica chymica.* Editions published 1608–1670 (and later).

Cumano, Marcello [Cumanus, Marcellus]. Quoted by Welsch's *Sylloge curationum et observationum medicinalium* (see below).

Dekkers, Frederick. In Barbette's *Praxim Barbettianam cum notis et observationibus Frederici Dekkers* (see above).

Deodatus, Claudius. *Pantheum hygiasticum.* Editions published 1628 & 1629.

Despars, Jacques [Partibus, Jacobus de]. Physician and Chancellor of the University of Paris; lived c. 1380–1458 (CERL).

Deuteronomy. [A book of the Bible].

Diemerbroeck. Probably Ysbrand van Diemerbroeck; lived 1609–1674. Wrote on human anatomy and on the plague (CERL).

Diogenes Laërtius. *Vita Sua,* books IV, IX. Third century biographer of the Greek philosophers.

Dionysius of Halicarnassus. Book VIII.

Dionysius the Aeropagite. *De divinis nominibus.*

Dioscorides. *De materia medica.*

Dioscorides. *"De re medica".*

Dodoens, Rembert [Dodonaeus]. Dutch botanist and Professor of Medicine at Leiden; lived 1517–1585. (CERL).

Döring, Michael. *"De peste"* possibly *Fasciculus quorundam tractatuum de peste.* 1631, Breslau.

Dubois, Jacques [Sylvius, Jacobus]. French physician and anatomist; lived 1478–1555 (CERL).

Duchesne, Joseph [Quercetanus]. *"De affectibus capitis"* possibly *Tetras gravissimorum totius capitis affectuum*. Editions published 1606–1617, Marburg.

Duchesne, Joseph [Quercetanus]. *Pharmacopoea dogmaticorum restituta*. Editions published 1607–1631.

Durante, Castore. Italian botanist and physician; lived 1529–1590 (CERL).

Ecclesiastes. [A book of the Bible].

Elideo de Padoani [Paduanis, Helidaeus de.] Quoted in the letters collected by Scholz, see below.

Englehart, Dr. Chief Physician in Anhalt (Hertodt, 1671).

Eobanus [Helius Eobanus Hessus]. Book III, *Heroidum*.

Epifanio, Ferdinando [Epiphanius, Ferdinandus]. Italian physician & historian; lived 1569–1638 (CERL). Cited by Hertodt as the author of *Empyrica* (but see next entry).

Epifanio, Nicola [Epiphanius, Nicolaus]. *Empyrica*. Editions published 1615 & 1625.

Epistle to the Hebrews. [A book of the Bible].

Ettmüller, Michael. Practitioner in Leipzig (Hertodt, 1671) and member of the *Academiae Naturae Curiosorum* (Anon, 1676a). Provided information by personal communication to Hertodt.

Evagrius. Evagrius Ponticus. Christian mystic; lived 345–399 AD.

Fabry von Hilden, Wilhelm [Fabricius Hildanus]. "*Observationes chirurgicae*" & "*Chirurgicae observationes*" possibly *Selectae observationes chirurgicae*. 1598. Or *Observationes et curationes chirurgicae*. Editions published 1606–1641.

Falloppio, Gabrielo [Fallopius, Gabrielis]. *De simplicibus medicamentis purgantibus*. 1565 & 1566, Venice.

Fehr, Johann Michael [Febrius, Michaelis]. "*Scorzonera Curiosorum*" possibly *Anchora sacra, vel scorzonera, ad normam & formam Academiae Naturae-Curiosorum*. 1666, Jena.

Fernel, Jean [Fernelius, Johannis]. *"Methodus medendi"* possibly *Therapeutices universalis seu medendi rationis*. 1607 & 1610, Hannover.

Ferrari, Giovanni Battista [Baptista] of Siena. *De florum cultura* [book IV]. Editions published 1633–1664.

Feynes, François [Feyneus]. *Medicina practica*. 1650, Lyon. [Also incorrectly cited as Feuneus by Hertodt and listed as *"Praxis medica"*, *"Practica"* and *"Practica medicina"*].

Finck, Joannes Vincentius [Finckius]. *Encheiridion dogmatico-hermeticum*. 1618 & 1628, Leipzig.

Finx, Erasmus [Francisci, Erasmus]. "*Ost- und West-Indischen lustgarten*" possibly *Ost- und West-Indischer wie ach Sinesischer lust- und staats-garten*. 1668, Nuremburg.

Flavius Vegetius. Ch. 24 *De bello* presumably from *De rei militari*.

Florenzola. Untraced.

Fonseca, Rodrigo da. *Consultationes medicae*. Editions published 1619–1625.

Foreest, Pieter van [Forestus, Petrus]. *Observationes et curationes medicinales*, book II. 1589, Leiden.

Fregoso [Fulgosus]. "Book I." Possibly Fregoso, Battista. Writer of biographies; lived 1453–1504 (CERL & WorldCat).

Freitag, Johann the elder. *Aurora medicorum Galeno-chymicorum*. 1630, Frankfurt.

Fuchs, Leonhart [Fuchius, Leonardus]. German physician and botanist; lived 1501–1566 (CERL).

Gabelkover, Wolfgang [Gabelchoverus]. *Curationum et observantionum medicinalium centuria* 1. Editions published 1611 & 1612.

Galen. "*Aphorisms*", the *Commentary on Hippocrates' Aphorisms*.

Galen. Book VI, *De simplicium medicamentorum facultate*.

Galen. *Commentary on Hippocrates' Epidemics*.

Galen. *De curanda ratione*.

Galen. *De diebus decretoriis*.

Galen. *Methodus medendi*.

Gans, Johann Ludwig [Gansius]. *Corallorum historia*. 1630 & 1669, Frankfurt.

Gattinara, Marco [Gatenaria]. "*Praxis medica*". Possibly a later title for *De curis egritudinum particularum noni Almansoris practica uberrima*.

Geiger, Malachias [Geigerus]. *Microcosmus hypochondriacus*. Editions published 1651 & 1652.

Gessner, Conrad. *Thesaurus Euonymi Philiatri*. Editions published 1552–1588. Incorrectly cited as by Ryff by Hertodt.

Glauber, Johann Rudolf [Glauberus]. *Miraculum mundi*. Editions published 1653–1660.

Gordon, Bernard de [Gordonius]. Author on hiccups. Physician; lived 1283–1320 (CERL).

Graba, Johann Andreas [Grabbae]. *Elaphographia*. 1667 & 1668, Jena.

Grégoire, Pierre [Gregorius Tholozanus, Petrus]. *De republica*. 1597 & 1609, Frankfurt.

Gregory the Great, St. *Moralia in Job*, book III.

Grembs, Franciscus Oswaldus [Grembsius]. *Arbor integra et ruinosus hominis*. Editions published 1657 & 1671.

Grüling, Philipp [Grylingius]. *Florilegium* [*Hippocrateo-Galeno*] *chymicum*. Editions published 1631–1665.

Grüling, Philipp [Grylingius]. *Medicinae practicae*. Editions published 1665 & 1668. Also probably the same as "*Praxis medica*" listed by Hertodt.

Hafenreffer, Samuel [Hafenrefferus]. "*De cutis affectibus*, book II", Latinized title of *Pandocheion aiolodermon*. 1630, Tübingen.

Hartmann, Johannes [Hartmannus]. *Praxis chimiatrica*. 1634.

Heer, Henri de [Heer, Henricus ab]. Physician; lived 1570–1636 (CERL).

Helmont [Helmontius]. Either Jan Baptista van Helmont, Flemish chemist and physician; lived 1577–1644 (CERL). Or his son, Franciscus Mercurius van Helmont, chemist and physician; lived 1614–1699 (CERL).

Heraclitus of Ephesus. Quoted by Diogenes Laërtius in his *Vita sua*, see above.

Heresbach, Conrad von [Heresbachius]. *De rei rustica*. Editions published 1570–1623.

Hermes Trismegistus. Purported author of the *Hermetic Corpus*.

Hertodt, Johann Ferdinand. *Tartaro-Mastix Moraviae*. 1669, Vienna.

Hertodt, Matthias Franz. Provided information by personal communication. (See also biography of Johann Ferdinand Hertodt, in the Introduction, section 1.1.)

Hiebner, Israel. *Mysterium sigillorum, herbarum et lapidum*. Editions published 1651 & 1653.

Hippocrates. *Aphorisms*.

Hippocrates. *De morbis III*.

Hippocrates. *Epidemics*.

Hoefer, Wolfgang [Hoeferus]. *Hercules medicus*. Editions published 1657 & 1665.

Hoffmann, Friedrich. "*Methodus medendi*", the *Opus de methodus medendi: iuxta seriem Wallaeianam, annexis fundamentis astrologicis, ex veterum ac recentiorum scriptis concinnatum/cum praefatione Johannis Michaelis*. 1668, Leipzig.

Honorius Pontifex [There were four popes plus one antipope named Honorius].

Horst, Gregor [Horstius]. Appendix *De variolis et morbillis*. 1606, Basel.

Horst, Gregor [Horstius]. *De tuenda sanitate*. Editions published 1615–1648.

Hortus Eystettensis, see Besler, Basilius.

Hotman, Antoine [Hottomannus]. "*Tractatus de dissolutione matrimonii ob frigiditatem*" Latinized title of *Traicté de la dissolution du mariage par l'impuissance & froideur de l'homme ou de la femme*. Editions published 1581–1656, Paris.

Houllier, Jacques [Hollerius]. *Commentarii in aphorismos Hippocratis*. 1613, Geneva.

Isaac Israeli. "*De dieta et herbis particularibus*" possibly *Dietae particulares*.

Isidore. *Etymologiae*, book XVII, *De rebus rusticis*.

Joël, François. *Practicae*, books I & II. 1617, Hamburg.

Joël, François. *Practicae*, books V–X. 1622, Lüneburg.

Jonston. Jan Jonston, naturalist and physician; lived 1603–1675 (CERL).

Jordan, Thomas [Jordanus, Thomas]. *De pestis phaenomena*. 1576, Frankfurt.

Jung, Georg Sebastian. Colleague of Hertodt, and practitioner in Vienna; supplied information by personal communication (Hertodt, 1671).

Justinian. "*Liber penitentialis*."

Justinian. *Novellae consitutiones*.

Juvenal. Roman satiric poet.

Keßler, Thomas [Keslerus]. "*Chymicus processus*" Latinized title of *Vierhundert außerlesene chymische proceß und stücklein*. 1632, Strasbourg.

Khunrath, Conrad. *Medullae destillatoriae et medicae*. Editions published 1614–1638.

Kircher [Kircherus]. Authors of this name, publishing before 1671 include: Athanasius Kircher, Christoph Kircher, Conrad Kircher and Johann Kircher (CERL).

Kornthauer, Job. "*Tractatus de peste*" possibly *De peste Aureoli Theophrasti Paracelsi tractatus*. Editions published 1613 & 1622.

Kyper, Albert [Kyperus, Albertus]. *Collegium medicum*. Editions published 1655 & 1666.

Lauremberg, Peter. *Apparatus plantarius*. Editions published 1632–1654.

Laurens, André du [Laurentium, Andreas]. *De crisibus*. Editions published 1596 & 1606.

Lejel, Tancred [Loelius]. *Curationes aliquot morborum rariores* (Extant in Bartholin's *Cista medica*, 1662 (see above)). Seventeenth century Danish physician based in Copenhagen.

Liberati, Liberato [Liberatis, Liberatus]. *Podagra politica*. Editions published 1637–1659.

Lipsius, Justus. *De amphiteatris quae extra Romam libellus*. 1604, Antwerp.

Livy [Livius]. *Third decade*.

Lobel, Matthias de [Lobelius]. *Stirpium observationes* in *Plantarum seu stirpium historia, 2 vols*, 1576.

Lobel, Matthias de [Lobelius]. *Stirpium adversaria nova*. 1570/71, London.

Locellus, Gulielmus. Inventor of a distilled water for cataracts (Hertodt, 1671).

Lotichius, Johann Peter. "*Dissertatio de febribus*" possibly *Paradoxon, sive de febribus in genere dissertatio theorico-practica*. 1627, Frankfurt.

"*Lugdunum.*" We have assumed that Hertodt used the same nickname for this book as English seventeenth-century botanist John Ray (see Oswald & Preston, 2011, p. 141), and that the proper title is: Dalechamps, Jacques. *Historia plantarum generalis*. 1578, Lyons.

Manardo (or Manardi), Giovanni [Manardus (listed incorrectly as Monardus in Hertodt's original text), Johannes]. Italian physician and author; lived 1462–1536 (CERL).

Marcellus. Untraced.

Marci, Jan Marek. Professor and physician in Prague (Hertodt, 1671); lived 1595–1667 (Gillespie, 1970–1990, IX, pp. 96–98).

Martial. *Liber spectaculorum*.

Mattioli, Pietro Andrea [Matthiolus]. "Book I."

Mattioli, Pietro Andrea [Matthiolus]. "*Herbarium*", the *Di Pedacio Dioscoride Anazarbeo libri cinque della historia, & materia medicinale tradotti in lingua uolgare italiana*. 1542, and later editions in Latin.

Mercato [Mercatus, Ludovicus]. *Praxis medica*. 1611, Venice.

Mermann von Schönburg, Thomas. German physician; lived 1559–1612 (CERL).

Mesue [Yuhanna ibn Masawaih]. "*De medicina benedicta*".

Michael, Johann [Michaelis, Johannes]. "*Medicinae oraculum*".

Minderer, Raymund [Mindererus]. *Aloedarium marocostinum*. Editions published 1616–1626.

Mirandola, Giovanni Francesco Pico della. "*Astrologia*, book X", *Disputationes adversus astrologos*. 1495, Venice.

Möbius, Gottfried. *Anatomia camphorae*. 1660, Jena.

Montagnana, Bartolomeo. *Consilia medica.* Editions published 1475–1564.

Monteux, Sebastien de [Montius (listed incorrectly at Montanus in Hertodt's original text), Sebastianus]. French physician; lived 1480–1599 (CERL).

Moreau, François [Moraeus Bruxiliensi, Franciscus]. *Opus de febre maligna noviter.* Editions published 1669 & 1670.

Morel, Pierre [Morellus, Petrus] "*De materia medica*" possibly *Methodus praescribendi formulae remediorum. Cum adjuncto materiae medicae systemate.* 1659, Amsterdam.

Müller, Philip [Mullerus, Philippus]. *Miracula et mysteria chymico-medica.* Editions published 1623 & 1656.

Mutzler, Konrad [Mutzlerus]. *Bezoardicum animale.* 1641.

Mylius, Johann Daniel. *Antidotarium medico-chymicum.* 1620, Frankfurt.

Mynsicht, Adrian von [Mynsichtus, Hadrianus von]. *Thesaurus et armamentarium medico-chymica.* Editions published 1638–1670.

Naevius, Johannes. German physician; lived 1499–1574 (CERL). Quoted in the letters collected by Scholz, see below.

Neucrantz, Zacharias. Wrote books on pleurisy and on arthritis published 1613 & 1614.

Nonius. Could be Nonius Marcellus, fourth-fifth century Roman grammarian, otherwise, untraced.

Nüremburg. "Ch. 49" [concerning an "Indian root", probably turmeric].

Ovid. *Ars amatoria.*

Ovid. *Metamorphoses.*

Ovid. *Tristia.*

Pachymere, George [Pachymerus, Gregorius]. "*Historiae*, book VIII."

Padovani, Fabrizio [Paduanus, Fabricius]. Italian philiopher and physician, active 1605 (CERL). Cited by Hertodt as the author of "*De anno climacterico*" but see under Selvatico, Giovanni Battista.

Paracelsus [Philippus Theophrastus Aureolus Bombastus von Hohenheim]. Physician, alchemist and pioneer of new ideas; lived c. 1493–1541.

Paré [Paraeus]. Ambroise Paré, French physician; lived 1509–1590 (CERL).

Paul of Aegina [Paulus Aegineta]. Seventh century Byzantine Greek physician.

Paulli, Simon. *Quadripartitum botanicum* Editions published 1666–1668.

Penot, Bernard. French physician and alchemist; lived 1520–1617 (CERL). Quoted in Petri's *Asylum languentium* (see below).

Petrarch. *De aegritudine tibiarum. Dialogus XCV.*

Petrarch. *De remediis utriusque fortunae.*

Petri, Georg Christoph. *Asylum languentium, sive carduus sanctus vulgo benedictus.* 1669, Jena.

Piemontese, Alessio [Pedemontanus, Alexis; a pseudonym for Girolamo Ruscelli]. "*De secretis*" Original edition *De' Secreti.* 1557, Venice. Later editions in many European languages.

Piens, Franciscus. Wrote on fevers. Physician; lived 1600–1699 (CERL).

Pierio. "Volume VI, book XXXII." See next.

Pierio Valeriano, Giovan Pietro [Pierius Valerianus, Johannis]. Italian humanist, historian etc.; lived 1477–1560 (CERL).

Pincier, Johann [Pincierus, Johannes]. *Aenigmata*. Editions published 1605 & 1655.

Platter, Felix. *Praxis medica*. 1625, Basel.

Plautus, Titus Maccius. *Captivi* ["the Captives"].

Pliny. *Historia naturalis*.

Plotinus. *Enneads*.

Plutarch. *De consolatione ad Apollonius* [Hertodt incorrectly assigned this work to Seneca].

Plutarch. *Pyrrhus*.

Polemann, Joachim [Polemannus]. *Novem lumen medicum de mysterio sulphuris philosophorum* in: *Theatrum chemicum* Vol. 6, pp. 600–675. 1661, Strasbourg.

Popp, Johann [Poppius, Johannes]. German physician and chemist; born 1577 (CERL). Quoted by Gans in his *Corallorum historia*, see above.

Potier, Pierre [Poterius, Petrus] "*Observationum & curationum centuria* 1" possibly *Insignium curationum, et singularum observationum centuria prima*. Editions published 1615–1624.

Potier, Pierre [Poterius, Petrus] "*Observationum et curationum, centuria* 3."

Potier, Pierre [Poterius, Petrus]. *Pharmacopoea spagyrica*. Editions published 1624 & 1635.

Pratensis, Jason. *De cerebri morbis*. 1549, Basel.

Prévost, Jean [Praevotius, Johannes]. *Selectoria remedia multiplici comprobata: libellus posthumus*. 1659, Frankfurt.

Procopius. *De bello Persico*.

Publilius Syrus. Latin writer; lived 85–43 BC.

Publius. Untraced. Possibly a misspelling of Publilius.

Pythagoras. "*De divinis*." Probably cited from a doxographical source (Diogenes Laërtius; Sextus Empiricus, Aetius).

Quintus Curtius Rufus. First century Roman historian.

Remmelin, Johann. *Catoptrum microcosmici*. Editions published 1619–1660.

Renou, Jean de [Renodaeus]. "*Commentarius*". French physician; lived 1568–1620 (CERL).

Rhazes [Abū Bakr Muhammad ibn Zakariyyā al-Rāzī]. Persian physician and alchemist; lived 854–925 AD.

Rhumel [Rhumelius]. Possibly Johann Conrad Rhumel, physician; lived 1597–1661 or 1574–1630 (CERL): or Johann Pharamund Rhumel, German alchemist & physician; fl. 1620–1645 (CERL).

Rivière, Lazarus [Riverius]. "*De suffusione*".

Rivière, Lazarus [Riverius]. *Institutiones medicae: in v libros distinctae*. 1662, The Hague.

Rivière, Lazarus [Riverius]. *Praxis medica*. Editions published 1640–1664.

Roderigo a Castro, Esteban [Roderici Castrensis, Stephani]. *Medicae consultationes*. 1639–1644, Florence. Also cited as "*Consultationes medicae*" by Hertodt.

Roet, Isaac. *Pestis adumbrata*. 1666, London.

Rolfinck, Werner [Rolfincus]. *"Specialis methodus medicinae"* & *"Methodus medicinae"*, *"Methodus medicinae specialis"* possibly *Ordo et methodus medicinae specialis*. Editions published 1660–1671.

Rolfinck, Werner [Rolfincus]. *Chymia in artis formam redacta*. 1641, Jena.

Rolfinck, Werner [Rolfincus]. *Consultatio medica de mania*. 1666, Jena.

Rolfinck, Werner [Rolfincus]. *De vegetabilibus, plantis, suffruticibus, fructibus, arboribus in genere*. Editions published 1669 & 1670.

Roy, Hendrik de [Regius, Henricus]. "*Consilia medica*." Dutch physician and Professor of Medicine at Utrecht; lived 1598–1679 (CERL).

Ruel, Jean [Ruellius]. *De natura stirpium*. Editions published 1536–1543.

Ruland, Johann David [Rulandus]. *Pharmacopoea nova*. 1644, Nuremburg.

Ryff, Peter [Ryffius]. Swiss mathematician, physician and Professor at Basel; lived 1552–1629. (CERL). Hertodt wrongly lists him as the author of *Thesaurus Euonymi*. (See Gessner, Conrad, above).

Sachs von Löwenheim, Philip Jakob. *Gammarologica*. 1664 & 1665, Frankfurt.

Sachs, Philip Jakob. *"De usu praeparationis croci."*

Sala, Angelo [Salae]. *Essentiarum vegetabilium anatome*. 1630 & 1635, Rostock.

Sala, Angelo [Salae]. *"De peste"* possibly *Tractatus de peste*. Editions published 1617 (in French) & 1641 (in Latin).

Sallustius. Fourth century Roman writer.

Scala, Giulio Cesare Della [Scaliger, Julius Ceasar]. *"Ad Cardanum excercitationes", "De subtilitate ad Cardanum"* & *"Excitatio"* possibly *Exotericarum excertationum liber xv, de subtilitate ad Hieronymum Cardanum*. Editions published 1557–1634.

Schenck, Johannes [Schenkius]. *Observationes medicae*. Editions published 1587–1665.

Scholz, Lorenz [Scholzius, Laurentius]. *Consiliorum medicinalium*. Editions published 1598 & 1610.

Schönborn, Samuel. *Manuale medicinae practicae Galenico-chymicae*. Editions published 1637 & 1657.

Schwenckfeld, Caspar. *Theriotropheum Silesiae*. 1603, Liegnitz.

Selvatico, Giovanni Battista [Silvaticus, Johannes Baptista]. *De anno climacterico. Tractatus*. 1615, Pavia. [Hertodt incorrectly assigned this work to Fabrizio Padovani].

Seneca. *De consolatione ad Marciam*.

Seneca. *Moral epistles*.

Seneca. *Tragedies*.

Sennert, Daniel [Sennertus]. *"De febribus"* possibly *Epitome liborum de febribus*. 1655, Amsterdam. Also the author of other titles on fever.

Sennert, Daniel [Sennertus]. *Practica medicina*. Editions published 1648–1654, Wittenberg.

Sennert, Daniel [Sennertus]. *Auctarium epitomes physicae*. Editions published 1635–1651.

Serapion. *De simplici medicina*.

Sethus, Simon [Simeon Seth]. Eleventh century Jewish Byzantine physician.

Settala, Luigi [Septalius, Ludovicus; Septal, Louis]. "Book IX." Professor of Philosophy and Medicine at Milan; lived 1552–1633 (CERL).

Sextus. Sextus Empiricus, physician & philosopher.

Sorbait, Paul de. Dean of the Medical Faculty in Vienna, and Court-Physician of Her Majesty Empress Eleonora; lived 1624–1691. (Hertodt, 1671; CERL).

Spenholz, Andreas. *"De Peste"* possibly *Kurtzer und klarer bericht von der natur und aigenschafft der grawsamen abschewlichen tyranischen seucht der pestilentz*. 1639, Linz.

Stocker, Johannes [Stockerus]. *Empirica*. 1601, Frankfurt.

Suetonius [Gaius Suetonius Tranquillus]. *De vita Caesarum*. (Now known as *The twelve Caesars*).

Tabernaemontanus, Jacobus Theodorus [Theodorus aus Bergzabern, Jakob]. "Book II." German botanist; lived 1522–1590 (CERL).

Tentzel, Andreas [Tentzelius (listed incorrectly as Tenzelius in Hertodt's originaltext)]. "Tract on the plague" possibly *Philosophus-medicus de Peste*. 1627, ?Nordhausen.

Teodosi, Giovanni Battista [Theodosius, Joannes Baptista]. Italian physician; lived 1475–1538 (CERL).

Teuber, Johannes [Teuberus]. *Arthritis*. 1653, Prague.

Theatrum vitae humanae, see Zwinger, Theodor.

Themistius. Greek philosopher; lived 317–c. 390 AD.

Theophrastus. Greek philosopher and botanist; lived 371–c. 287 BC.

Thucydides. "Book III." [Of the *History of the Peloponnesian war*].

Tilemann, Johannes [Tilemannus]. *Appendix de materia medica*. Editions published 1650–1666.

Timaeus, Balthasar von Güldenklee. *Casus medicinales praxi triginta sex annorum observati*. 1662 & 1667, Leipzig.

Timaeus, Balthasar von Güldenklee. "Book I, *De affectibus capitis*."

Timaeus, Balthasar von Güldenklee *"Descriptiones medicamentorum epistolae"* & *"Epistolae medicinales"* & *"Eipstolae"* possibly *B. Timaei von Güldenklee et quorundam Germaniae, Galliae et Italiae medicorum epistolae et consilia*. 1665, Leipzig. Or *Epistolae et consiliae*. 1665, Leipzig.

Tulp, Nicolaas [Tulpius, Nicolaus]. *Observationes medicae.* 1652, Amsterdam.

Turquet de Mayerne, Théodore [Turquetus]. Swiss physician and chemist; lived 1573–1655 (CERL).

Untzner, Matthias [Untznerus (listed incorrectly as Unznerus in Hertodt's original text)]. *Antidotarium pestilentiale in duos libros tributum.* 1621, Halle an der Saale.

Valle, Rolando Della [Capellutius, Rolandus]. "*Tractatus chirurgicae de peste*" possibly *De curatione pestis tractatus.* 1648, Braunschweig.

Vallès, Francisco [Vallesius, Franciscus]. Spanish physician; lived 1524–1592 (CERL).

Vettori of Faenza, Benedetto [Victorius Faventinus, Benedictus]. Professor of Medicine at Bologna; lived 1481–1561 (CERL).

Vigo, Giovanni da [Vigonio]. Italian surgeon; lived c. 1450–c. 1525 (CERL).

Virgil. *Georgics IV.*

Wecker, Johann Jakob [Weckerus]. *De secretis.* 1582 and later editions.

Welsch, Georg Hieronymus [Welschius]. *Sylloge curationum et observationum medicinalium.* Editions published 1657–1668.

Widemann, Karl [Widemannus, Carolus]. "*Von der pest*" & "*Tractatus de peste*" possibly *Drey außerlesene tractat von der pest* by Paracelsus, Rolando Della Valle & Karl Widemann. 1640, Frankfurt.

Willich, Jodocus [Willichius]. "*Consilia medica*" possibly from one of several books on the plague by this author. German physician, Professor at Frankfurt; lived 1486–1552. Or his son: also a physician; lived 1543– (CERL).

Willis, Thomas. "*De febribus*" possibly *Diatribae duae medico-philosophicae ... altera De febribus ...* 1562–1669 (and later editions).

Wirbunk, Johannes. "*Artzney-buch.*" Hertodt may have mistaken the author's name, and the book in question is possibly instead by Johann Wier (*Artzney-buch,* editions published 1580–1599, Frankfurt).

Wirsung, Christoph. "Part 2, ch. 5". German physician; lived 1500–1571 (CERL).

Wittich, Johann [Wittichius]. "*Consilia Germanica*" & "*Consilia medicinalia*" possibly *Nobilissiorum ac doctissimorum Germaniae medicorum consilia.* 1604, Leipzig.

Zacuth of Lisbon [Zacutus; Zacuto, Abraham]. "*Pharmacopoea*" possibly part of his *Praxis historiarum.*

Zacuth of Lisbon [Zacutus; Zacuto, Abraham]. *Praxis historiarum.* Editions published 1643–1667.

Zacuth of Lisbon [Zacutus; Zacuto, Abraham]. *Praxis medica admiranda.* Editions published 1643–1667.

Zecchio, Giovanni [Zecchius]. *Consultationes medicinales.* Editions published 1599–1650.

Zwelfer, Johann [Zwelferus]. *Pharmacopoeia Augustana reformata.* 1653, Gouda.

Zwelfer, Johann [Zwelferus]. *Pharmacopoeia regia.* Editions published 1652–1668.

Zwinger, Theodor (& Lycosthenes, Konrad (& others in later editions)). *Theatrum vitae humanae, omnium fere eorum, quae in hominem cadere possunt bonorum atque malorum, exempla historica, ethicae philosophiae praeceptis accommodata, et in* XIX *libros digesta.* Editions published 1565–1666 (and later).

APPENDIX 2

Animadversion on Hertodt's *Crocologia* by Wenzel Maximilian Ardensbach (1671)

ANIMADVERSION

ON

HERTODT'S

CROCOLOGIA

by

WENZEL MAXIMILIAN

ARDENSBACH

PhD & MD

1671

The named author published a third short work; whether this is more profound than the former works, or more superficial than the same, should be left open to judgement.

Doubtless, I respect *Tartaro-Mastix* more than *Crocologia*. For, in the former, the author bothers to present us something new, whereas, in the latter, he puts together in one block all the things already said by others in the past. If you take out the things written by others, little would be left to the author. However, I do not reproach the work, as you will find in one book what would need to be sought in several books. Nevertheless, it seems to go too far, that, in order to show that saffron cures all diseases, he exclusively brings together compound medicines of such kind that only contain a very little quantity of saffron, not taking into consideration whether that medicament is tested by experience, it suffices only that an author has mentioned it. Let the one who wishes try, and see [what happens].

But of what use is it, whether that remedy has saffron mixed in it, if it is exceeded by great quantity by other medicaments appropriate for the disease? What [difference] do some grains of saffron make amongst several drachms of other ingredients? Since saffron is warm and dry according to the author, if mixed with cold and moist medicaments, then its action is overthrown or hindered.

In several places Zwelfer, in his remarks on the Augustan *Pharmacopoiea*, mocks and opposes such absurd denominations and appropriations of virtues. However, not only *Crocologia*, but also several works of others, have such absurdity and feebleness. It seems to me that there are writers of such kind like those who, while they drank water

 | DOI:10.1163/9789004435292_030

during the whole time of the meal, and only one little glass of wine, they pride themselves that they are wine experts and sated with wine. Doubtless, in several compound medicines listed in *Crocologia*, saffron is comparatively very insignificant, therefore what can it do, if one or two grains of it are taken?

In some medicines, saffron behaves like a cordial, added either because of its fragrance or because of its colour; in others it seems to be irrelevant, and, in several compound medicines, it could be omitted. Likewise, the rest could do without saffron the same as they are believed to do with it.

But, please, what is this way of writing books? Somebody thinks of a topic of writing, for instance, mummy. Whatever he finds about it, whether relevant or irrelevant, he throws it all in, and compiles it in one, so that a miraculous lump is made. All the diseases are presented, wherever mummy or the chosen topic have something to do with them. Therefore, remedies from several authors are sought, which contain it. If something beneficial is found, the medicament is exalted with praises; if this does not work in the practical application, then it should be blamed on the rest of the ingredients added, not on the chosen one, because the rest are stronger in quantity and action. Also, many authors of this kind have the very bad habit, that whatever they write that is beneficial for this or that disease, they exalt it with the highest praises. One time they say it is a panacea, another time that is a specific, another that it is a gift of God, then that it is totally tested, then that it is the hope of human health, so that you believe that it is a sacrilege if you omit such a medicament in the cure of some disease. But afterwards by experience you bewail very much that it has been prepared; you will be sorry that you have been deceived and perhaps that a patient died.

One [author] is followed by another, another copies from another. The first person devised the medicine in his head and gave it a nice name; the second, relying on the first, includes it in his work, the third copies it again and, thus, the fourth as well. The first has not tried the validity and effect of the medicament. He deceived the second, the second deceived the third, and so on. But now who suffers the pains of such thoughtlessness? The wretched patient, who does not benefit from such remedies, if not killed by them.

Now, the wise doctors should open their eyes and give their opinion as to how those who put together a book of such kind are useful to medicine or to humankind? Don't those who collect in one place from other books the things found described, not tested, not experimented, make a new dress from old rags? Surely it would be better to refrain from such a useless task, or if the itch to write books is too great, it would be better to immerse oneself into the experimenting of things, to try the devised medicaments, to explain with observations and not to write anything that the soundest explanation does not demonstrate or that experience does not prove incontestably.

This manner of writing is hard; it requires hard work. We would have fewer books, but better ones, such that they would be more beneficial for the patient. A nice title is given to the book, if you have bought it and read it through, you will inevitably be sorry for the money spent for nothing.

I do not say anything about what one should think about *Crocologia*; those who chewed it thoroughly should judge. Many irrelevant things are included. I believe that it is unseemly for a doctor to teach the way to catch mice and to make mouse-traps. The cultivation of saffron, because it can be attributed to botany, is just about more tolerable. The various medicaments, which are instructed to be made with saffron, I gladly confess, are easily procured, but I believe that nothing is added to saffron through the preparation because it is entirely pure and almost entirely essensificated. It is not unbefitting that saffron can be assumed in an inferior or more convenient form, prepared in this way or that. Nevertheless, some preparations give rise to doubts, and chemists will doubt that preparations of this kind are going to work in practice. The best judge will be practice and experience, which should be taken into account.

Most definitions of diseases are general, and dated, some new. I examined five questions related here too in *Tartaro-Clypeus*.[1] I have not yet had the time to turn upside down the others. I hope that soon a time will come, when I will recall into examination all the explanations and causes of the diseases, old as well as new, and I will demonstrate how each of those proposed in *Crocologia*, which I bring under scrutiny too, stands up to experiment. Meanwhile, it is pleasing to briefly run through the questions attached here.

QUESTION I
Is saffron a remedy against monkshood?

Pieter van Foreest may claim that it can be a remedy, however, there are not many others [who do so], see Sennert, Jonston, Macasius and others. Perhaps van Foreest believed that monkshood such like it grows in Poland is, as they say, also assumed harmlessly, then saffron will easily be a remedy. But I do not believe that Poles eat flowers of monkshood, because surely those are found poisonous everywhere.

If, however, it can be proved to be true, it should be ascribed inevitably to transplantation, as in the case of the peach-tree, certainly not to saffron. In fact, Poles do not consume one ounce of saffron at a time in broth. Any sensible person would know what happens if one ounce is put into a measure of water: that many saffron threads would remain undissolved,[2] and the broth could not be consumed for the bitterness,

1 This book is by Ardensbach. The full title is *Tartaro Clypeus, excipiens Tartaro Mastigem Hertodianum*. 1671. Prague.

2 Modern recipes often still use the word "dissolve" to describe what happens to saffron threads as they give up their colour when in a liquid. A more appropriate word is "infuse." When talking about the use of an ounce of saffron, Ardensbach is describing a saturated solution, where

and, according to Schröder and others, if three drachms kill, even more so does one ounce or eight drachms. If now, then, eight drachms are added to monkshood for correction, then poison is added to poison; is this an antidote against poison? Since saffron kills, it is poison.

QUESTION II
Does saffron prevent or cause intoxication?

Saffron is an anodyne, they say; saffron induces sleep, they claim. Wine too induces sleep, numbs pain when it induces sleep, if it does not prevent wakefulness one must not expect this from the other either. This is limited reasoning. The experience of Dr Michael demonstrates that daily saffron-flavoured broths or wine inebriate very much. Then, how is saffron going to prevent intoxication? If someone drowned in wine is prone to sleep, and a soporific is administered to him, will he stay awake? This is clearly the point.

QUESTION III
Does saffron induce sleep or wakefulness?

Saffron, with abundant sulphur anodyne, is used by the author and by others to induce sleep, which experience confirms. However, the author wishes to say that it induces wakefulness too, and, thus, that it drives away the sleep by the reason that saffron is dry and hot, and to a certain degree it acts against sleep-inducing conditions, moist as well as cold. Now, I argue this: saffron prevents sleep because it is dry and hot, therefore, if it prevents wakefulness, it will be cold and moist. In consequence, it is demonstrated that, if sleep-inducing conditions are cold and moist as he says, then wakefulness-inducing conditions, in contrast with the former, will be hot and dry, then if saffron prevents wakefulness it will be cold and moist, because opposites are removed by opposites. Saffron will then be hot, cold, moist, dry, and this indeed in the extreme degree. But are opposites in the extreme degree compatible with each other? If you allege hidden qualities, you will hear it said by sensible doctors that these qualities are ignorance made manifest.

QUESTION IV
Does saffron tinge a foetus in the womb?

no more crocin can infuse into the liquid. In this situation, the excess saffron threads are effectively wasted. Less saffron could have been used to give the same effect.

A man of exceptional intelligence, my old teacher, Dr Marek Marci, in his *Philosophia restituta*, claims that saffron enters the womb and tinges the waters.[3] However, ingested saffron does not produce a yellow stain on any foetus because it tinges the foetus only as far as the surface and the yellow colour can be very easily washed off, as experience teaches. If, in the anatomy of the dog, the blood and lacteal vessels do not become yellow, how can the colour arrive invulnerable into the womb? If the author solved this matter, the others and I would value *Crocologia, Tartaro-Mastix* and *Opus mirificum diei sextae* more. However, he tried to lay down various reasons, but he demonstrated nothing; because he himself has doubts about the ascribed reasons, he claims that it can be this or that reason. Another will say that it is neither of them. I, however, do not have time now to solve the difficulty. With time I will make amends of what has been omitted here.

THE END

3 Marek Marci, Jan. *Pan en pantōn, seu philosophia vetus restituta.* 1662, Prague.

APPENDIX 3

Glossary

4 greater cooling seeds, n. The seeds of lemons, gourds, melons and cucumbers (Culpeper, 1683).

5 precious stones, n. Sapphire, garnet, emerald, hyacinth (q.v.) and carnelian (*Lexicon Pharmaceuticum*, 1694, p. 325).

Acacia, n. Inspissated (q.v.) juice pressed from unripe pods of a leguminous tree, thought to be from the same plant as gum Arabic (q.v.) (Peachey, 1707; James, 1747), or the juice of sloes (*Prunus spinosa* L. [Rosaceae]) (Blancard, 1708).

Acovistic, adj. Assumed meaning: helping the hearing ("*Oleum acovisticum* oil to help hearing", Quincy, 1722, p. 590).

Aegilops, n. Ulcer of the inner canthus of the eye; a lacrimal fistula.

Agaric, n. (white agaric) Nowadays the word "agaric" is a general term for a fungal fruiting body with a distinct pileus (cap) supported by a stipe (stalk) and with gills underneath. Additionally it may refer to the red-with-white-spots toadstool *Amanita muscaria* (fly agaric). However, in historical pharmacy, "agaric" meant the unrelated polypore bracket-fungus of larch trees (and other conifers), *Laricifomes officinalis* (Vill.) Kotl. & Pouzar (syn. *Polyporus officinalis*).

Agrimony, n. The plant *Agrimonia eupatorium* L. [Rosacece]. Old World.

Affection, n. Abnormal bodily state; a disease; a medical complaint or condition.

Affuse, v. To pour onto something.

Alabastrine, adj. Of or pertaining to alabaster.

Albugo, n. A whitish spot on the cornea; a corneal opacity.

Alembic, n. Broad definition: historical apparatus for distilling, but more properly the top part of such apparatus (q.v. cucurbit).

Alexiteric, adj. & n. Protecting against a poison; a medicine with this property. Historical alternative to alexipharmic (though Blancard (1708) defines alexiteric as a preservative against poision and alexipharmic as an expeller of poison).

Alkool, n. A very fine powder, often produced by rubbing the ingredients on a marble slab (Tomlinson, 1657, book III, p. 107). The OED spells this word "alcohol", but we have retained the older spelling to avoid confusion with the modern definitions of "alcohol".

Allotriploid, n. An organism with three sets of chromosomes derived from different parent species.

Aloëphanginae, pills of, n. A kind of pill made with aloes (q.v.) and other ingredients (Culpeper, 1683, p. 180).

 | DOI:10.1163/9789004435292_031

Aloes, n. The dried leaf-sap of succulent species of *Aloe* [Asphodelaceae]. The different kinds in *Crocologia* were named for their origin or appearance: (i) **Socotrine aloes** (Succotrine aloes) from *Aloe perryi* Bak. of Socotra, an island and an archipelago in the Arabian Sea; (ii) **bright aloes** (*aloes lucida*), a clear and transparent form; or (iii) **hepatic aloes** (*aloes hepatica*) with the consistency and colour of liver, and a stronger smell than the Socotrine kind (Pechey, 1707).

Aloes, rosated, n. (*aloes rosata*) A preparation of aloes treated with oil of roses (Culpeper, 1683, p. 181).

Aloes-wood, n. (lign-aloes, agarwood) Highly-scented heartwood of *Aquilaria malaccensis* Lam. (syn. *A. agallocha*) [Thymelaceae] of Indomalasia, and related species (Ng et al., 1997), resulting from fungal infection of the trees. Unrelated to aloes, above. Aloes-wood is one of the most expensive natural substances and *Aquilaria* spp. are listed by CITES Appendix II. The source of oil of oud used in perfumery.

Alypta moschata, n. The name of a kind to troche (q.v.) (Tomlinson, 1657, p. 634; Culpeper, 1683, p. 193).

Ambergris, n. Aromatic grey susbstance produced in the bile ducts of sperm whales, *Physeter macrocephalus* L. and later evacuated into the sea.

Ammoniac, gum, n. (ammoniacum) Gum from *Ferula ammoniacum* (D.Don.) Spalik et al. [Umbelliferae] (syn. *Dorema ammoniacum*) of Iran to India, produced in response to puncture wounds by a species of beetle (Flückiger & Hanbury, 1874).

Anasarca, n. A dropsical affection of the subcutaneous cellular tissue of a limb or other large surface of the body.

Anodyne, adj. & n. Easing pain; a medicine with this property.

Anthophyllos, n. A product of the clove tree, *Syzygium aromaticum* (L.) Merr. & Perry [Myrtaceae], either: (i) the "refuse of cloves" (Mosan, 1605), possibly tiny pieces of the sepals or petals of the clove-buds that become dislodged in the drying and handling of the spice. Or (ii) dried, flowered cloves (Motherby & Wallis, 1755). Or (iii) dried fruits of the clove tree (Flückiger & Hanbury, 1874). (Cloves themselves are the dried flower buds *before opening*).

Anti-hectic, Potier's, n. A preparation containing tin (Hill, 1751, p. 46) to alleviate a hectic fever.

Antimony, diaphoretic, n. (butter of antimony) A diaphoretic (q.v.) substance made by heating antimony and nitre in a crucible (Motherby & Wallis, 1745).

Anti-nociceptive, adj. Alleviating pain, especially that associated with tissue damage.

Anti-phthisic, adj. & n. Treating or alleviating phthsis; a medicine with this property.

Anti-tussive, adj. Alleviating coughs.

Anxiolytic, adj. Reducing anxiety.

Aperient, adj. & n. Opening the bowels; a medicine with this property. In some cases the word also refers to an opening of the blood vessels.

Aperient roots, n. According to Quincy (1719, p. 65) the five greater opening/aperient roots are: smallage (q.v.), fennel, asparagus, parsley, butcher's broom (*Ruscus aculeatus* L. [Asparagaceae]). The five lesser opening roots are: grass (couch-grass: *Elymus repens* (l.) Gould), madder (q.v.), eryngo (sea-holly roots: *Eryingium maritimum* L. [Umbelliferae]), capers (*Capparis spinosa* L. [Capparaceae]), "cammock" (restharrow (Parkinson, 1640, ch. 18, p. 995): *Ononis spinosa* L. [Leguminoseae]).

Aphonia, n. The inability to produce voiced sounds; the loss of the voice.

Aposteme, n. (apostem) A large or deep-seated abscess or pustulent swelling.

Apozem, n. A decoction or infusion.

Aqua benedicta, Ruland's, n. Synonymous with *vinum antimonii* and *vinum stibiatum*. A preparation of antimony (Wollkopf, 1811).

Arcanum, n. Literal meaning: a secret. In medicine an arcanum was one of the great secrets sought by alchemists, i.e. a marvellous remedy or elixir.

Arcanum duplicatum, n. ["double secret"] A preparation (also called the salt of wisdom) of vitriol and nitre (Partridge, 1682, p. 13); an obsolete name for potassium sulphate (Mayne, 1860).

Archeus, n. The chief principle or power devised by Helmont to supervise the "animal economy" and preserve life. The chief archeus was thought to reside in the stomach (OED).

Argillous, adj. Clayey.

Asafoetida, n. Gum from *Ferula assa-foetida* L. of W Iran, *F. foetida* (Bunge) Regel of E Iran and *F. narthex* Bioss. of Afghanistan. Both plants are of the Umbelliferae.

Asarabacca, n. The plant *Asarum europaeum* L. [Aristolochiaceae]. N temperate regions.

Asphodel, n. The plants *Asphodelus albus* Mill. & *A. ramosus* L. [Asphodelaceae]. Mediterranean region.

Assume, v. To take into the body (food, nourishment etc.).

Atrabilarious, adj. Of or pertaining to black bile.

Autotriploid, n. An organism with three sets of chromosomes derived from a single parent species.

Avens, n. (wood avens, herb Bennet) The plant *Geum urbanum* L. [Rosaceae]. Europe, W Asia, Mediterranean region.

Balaustine, n. Large, dried, double flower of (wild) pomegranates *Punica granatum* L. [Lythraceae] (Miller, 1722).

Balm, n. (lemon balm) The plant *Melissa officinalis* L. [Labiatae]. S Europe.

Balsam of Peru, n. Product of the tree *Myroxylon balsamum* (L.) Harms var. *pareirae* (Royle) Harms [Leguminosae]. Venezuela-Peru.

Balsam, n. Product of *Commiphora gileadensis* (L.) C.Chr. [Burseraceae]. SW Arabia, Somalia & Sudan.

Balsamus sulphuris, n. A preparation of sulphur and oil (Mayne, 1860).

Basilisk, n. A mythological creature said to be king of all serpents, and able to cause death with a single glance.

Bdellium, n. Resin of various *Commiphora* spp. [Burseraceae]. Africa & Arabia.

Behen, n. (ben, been, ben root) Available in two broad forms, but with some disagreement over the species involved: (i) **red behen**, either "red valerian" (Culpeper, 1683), possibly *Centranthus ruber* (L.) DC [Caprifoliacaeae], or alternatively, the root of sea lavender (Miller, 1722; Motherby & Wallis, 1755), *Limonium binervosum* (G.E. Sm.) Salmon agg. [Plumbaginaceae]; (ii) **white behen**, "white valerian" (Culpeper, 1683): white flowered forms of either *Valeriana officinalis* L. or *Centranthus ruber* (L.) DC [both Caprifoliacaeae]. Other sources state **true white behen** is *Centaurea behen* L. [Compositae] (Mayne, 1860) and **common white behen** is bladder campion, *Silene vulgaris* (Moench) Garke [Caryophyllaceae] (Motherby & Wallis, 1755).

Benedicta laxativa, n. ["blessed laxative"] A particular preparation containing turpeth (q.v.) and other ingredients (Quincy, 1733, part III, p. 446).

Benjamin, gum, n. (gum benzoin) Gum from the tree *Styrax benzoin* Dryander [Styracaceae]. Sumatra.

Betony, n. The plant *Stachys officinalis* (L.) Trev. [Labiatae]. Eurasia.

Bezoar, n. A hard, stone-like concretion found in the stomach of animals, thought to be a counter-poison or antidote.

Bilious, adj. Of or pertaining to bile.

Birthwort, n. Medicinal plants of the species *Aristolochia* [Aristolochiacaeae], including: (i) **long** or **long-rooted birthwort**, *A. fontanesii* Boiss. & Reut. (syn. *A. longa*) (Mayne, 1860); and (ii) **round birthwort**, *A. rotunda* L. (Mayne, 1860).

Bishop-weed, n. Certain umbelliferous plants with small seeds, including: (i) **common bishop-weed**, *Ammi majus* L. of Mediterranean & W Asia, and *Sison amomum* L. Europe & Mediterranean region (Flückiger & Hanbury, 1874); or (ii) **true bishop-weed** (bishop's weed of Candy, true ammi, ajowan), *Trachyspermum ammi* (L.) Sprague (syn. *Ammi copticum*) originally from Egypt & Ethiopia.

Bistort, n. The plant *Bistorta officinalis* Delarbre [Polygonaceae] (syn. *Polygonum bistorta*). Temperate regions.

Bitter-sweet, n. (woody nightshade) The plant *Solanum dulcamara* L. [Solanaceae]. Eurasia.

Bitter-vetch, n. The plant *Vicia ervilia* (L.) Willd. [Leguminosae]. Mediterranean region to Afghanistan.

Bleak, n. A small, silvery, European freshwater fish, *Alburnus alburnus* L., of the family Cyprinidae.

Bloodstone, n. A red form of haematite, Fe_2O_3, and not that semi-precious stone, also called heliotrope, known as bloodstone today.

Bole, n. A type of earth that, unlike clay, easily dissipates in water, and remains in suspension (Hill, 1751).

Bole armeniac, n. A kind of earth from Armenia, rich in iron oxide (Hill, 1751).

Bombace, n. Cotton-wadding.

Bruchus, n. Either a locust, or also other insects including praying mantises and crickets (see diagrams in Aldrovandi, 1602, p. 413 et seq.). The name *Bruchus* was later assigned to a genus of beetles by Linnaeus (1767).

Bryony, n. Either of two unrelated climbing plants, both with large roots: (i) **black bryony, n.** *Dioscorea communis* (L.) Caddick & Wilkin [Dioscoreaceae] (syn. *Tamus communis*) of Europe & Mediterranean region; or (ii) **white bryony, n.** *Bryonia alba* L. [Cucurbitaceae] of Eurasia & Mediterreanean region. White byrony was used to prepare faeculae (q.v.) of bryony (Pechey, 1707; Miller, 1722).

Bugloss, n. The plant *Anchusa officinalis* L. [Boraginaceae]. Europe & Mediterranean region.

Burdock, n. The plant *Arctium lappa* L. [Compositae]. Temperate Europe.

Burnet, n. The plant *Sanguisorba officinalis* L. [Rosaceae]. Eurasia.

Butterbur, n. The plant *Petasites frigidus* (L.) Fries. [Compositae]. Europe.

Cachexia, n. A weakness and wasting of the body due to chronic illness.

Calamus scriptorius, n. A structure resembling a pen-nib, in the fourth ventricle of the human brain.

Calcine, v. To roast or burn something in a fire, reducing it to friable ashes.

Camphor, n. A substance derived from various trees: (i) **Chinese camphor** *Cinnamomum camphora* (L.) J. Presl. [Lauraceae] of China, Taiwan & Japan; (ii) **Sumatran camphor** *Dryobalanops aromatica* Geartn.f. [Diperocarpaceae] (Sumatran camphor was 100 times more expensive than the Chinese variety in 1563 (Flückiger & Hanbury, 1874)); or (iii) a **substitute** distilled from the root-bark of cinnamon trees (Pechey, 1707; Flückiger & Hanbury, 1874).

Canella, n. (white cinnamon) Inner bark of *Canella winterana* (L.) Geartn. (syn. *C. alba*) [Canellaceae] of Florida and the W Indies. Sometimes a syn. for cinnamon or cassia bark (q.v.).

Canthus, n. The outer or inner corner of the eye.

Caput mortuum, n. ["dead head"] The fixed matter left after all volatiles have been driven off a substance by e.g. distillation, sublimation.

Caranna, n. (gum caranna, caranna elemi) A resin brought from New Spain (Miller, 1722), or from "Carthage" (surely Cartagena in Colombia) (Pechey, 1707). Produced by plants of the Burseraceae including: (i) **American elemi**, *Bursera simaruba* (L.) Sarg. of S Florida, C America, W Indies; (ii) **West Indian elemi**, *Dacryodes excelsa*

Vahl (syn. *B acuminata*) of the W Indies; and (iii) **Carana elemi**, *Protium carana* (Humb.) Marchland of tropical America.

Carbuncle stone, n. A red precious stone, in early times being a ruby, garnet or spinel, but later being more restricted to a garnet.

Carminative, adj. & n. Dispelling trapped wind; a medicine with this property.

Carnation, clove, n. (gilliflower) Forms of *Dianthus caryophyllus* L. [Caryophyllaceae] with clove-scented flowers. Mediterranean.

Carob, n. Seed of *Ceratonia siliqua* L. [Leguminosae], introduced from Arabia to Iberia in the twelfth century.

Carpobalsam, n. Fruit of the balm of Gilead tree (Motherby & Wallis, 1755), *Commiphora gileadensis* (L.) C. Chr. [Burseraceeae] (syn. *C. opobalsamum*). SW Arabia, Ethiopia & Sudan.

Cassia fistula, n. (pudding-pipe) Fruit of *Cassia fistula* L. [Leguminosae]. Tropical Asia.

Cassia, n. (cassia bark, cassia lignea, Chinese cinnamon) The inner bark of *Cinnamomum aromaticum* Nees. [Lauraceae] (syn. *C. cassia*). Burma & China.

Castor-bean, n. (palma Christae) Seed of *Ricinus communis* L. [Euphorbiaceae]. NE Africa to the Middle East.

Castoreum, n. Aromatic, oily, substance from European and American beavers (*Castor fiber* L. and *C. canadensis* L. respectively).

Cataplasm, n. A poultice or plaster.

Catholicum, extract of, n. The name of a preparation thought of as a universal antidote (Tomlinson, part I, book II, p. 555).

Catmint, n. (catnip, nep) The plant *Nepeta cataria* L. [Labiatae]. Europe, SW & C Asia.

Celandine, n. Either of two unrelated plants: (i) **greater celandine**, *Chelidonium majus* L. [Papaveraceae] of temperate & sub-arctic Europe; or (ii) **lesser celandine**, *Ficaria verna* Hudson [Ranunculaceae] of Europe & W Asia.

Centaury, (lesser), n. The plant *Centaurium erythraea* Rafn. [Gentianaceae] (syn. *Chironia centaurium*). Europe.

Cerate, n. (cere-cloth) An external medicine, with a consistency between an ointment and a plaster.

Ceraunia stone, n. Thought originally to have fallen to earth in lightning strikes in thunderstorms, cerauniae eventually came to be recognised as ancient human artefacts (i.e. stone tools) in the seventeenth century (Goodrum, 2008).

Ceraunochrysos, n. (fulminating gold) An explosive precipitate made by reacting ammonia with gold chloride (Mayne, 1860).

Ceruse, n. (white lead) A lead compound with the formula $2PbCO_3{\cdot}Pb(OH)_2$.

Chaste tree, n. The plant *Vitex agnus-castus* L. [Labiatae]. S Europe.

Chervil, n. The plant *Anthriscus cereifolium* (L.) Hoffm. [Umbelliferae]. SE Europe to E Mediterranean region.

China root, n. The plant *Smilax china* L. [Smilacaceae]. E Asia.

Chyle, n. A milky fluid formed in the intestines; it is made of lymph fluid and digested fatty acids.

Chyliferous, adj. Containing chyle.

Chylous, adj. Of or pertaining to chyle.

Cinnabar, n. The mineral mercury (II) sulphide, HgS. Hill (1751) describes two kinds: native cinnabar, the natural mineral; and "factitious" or man-made cinnabar, and recommends the native form for internal use.

Cimmerian, adj. Perpetual darkness, named for a nomadic people of the ancient world who were believed to live in cloud and darkness.

Citron, n. Fruit of *Citrus medica* L. [Rutaceae]. N India. A lemon-like fruit with a thick rind.

Clavate, adj. Thickening or widening towards its apex.

Clyster, n. An enema; sometimes also a suppository. Also historically applied to medicines inserted into the ear canal.

Coarctation, n. A narrowing or constriction of the walls of a blood vessel, especially of the aorta.

Colature, n. The strainings through a sieve, not as refined as a filtrate.

Colcothar, n. A brown-red oxide of iron (Mayne, 1860).

Colliquament, n. The earliest stage of an embryo (OED). Motherby & Wallis (1745) define *colliquamentum* as an extremely transparent fluid in an egg.

Collyrium, n. A topical medicine for the eyes.

Colocynth, n. (coloquintida, bitter-apple) Gourd-like fruit of *Citrullus colocynthis* (L.) Schrad. [Cucurbitaceae]. Mediterranean & India.

Coltsfoot, n. The plant *Tussilago farfara* L. [Compositae]. Eurasia to Mediterranean region.

Columbine, n. The plant *Aquilegia vulgaris* L. [Ranunculaceae]. Europe.

Comfrey, n. The plant *Symphytum officinale* L. [Boraginaceae]. Eurasia.

Condite, n. & v. A preserve or pickle (or an electuary, q.v.); to preserve or pickle (OED); a simple medicine sweetened with honey or sugar (Blancard, 1708).

Confection, n. A medicine compounded of various drugs; a prepared state or condition.

Consound, n. (consolida) A name given to several herbs with reputations for healing wounds. In *Crocologia* "roots of both kinds of consound" are called for in one recipe. These could include: (i) **greater consound**, comfrey (Culpeper, 1683) (q.v.); (ii) **medium consound**, bugle *Ajuga reptans* L. of Europe (Culpeper, 1683), or upright bugloss, *A. pyramidalis* L. [both Labiatae] (Mayne, 1860); (iii) **lesser consound**, the daisy *Bellis perennis* L. [Compositae] (Culpeper, 1683) or self-heal (which see) (Mayne, 1860); or finally (iv) **consolida regalis**, larkspur (Culpeper, 1683) (q.v.).

Coral, red, n. *Corallium rubrum* L. [Coralliidae]. Mediterranean Sea, Straits of Gibraltar, around Cape Verde and Atlantic S of Portugal.

Coral, white, n. *Oculina* spp. Lamarck [Oculinidae] (Mayne, 1860). Gulf of Mexico & Caribbean waters.

Coralline, n. *Corallina officinalis* L. [Corallinaceae]. Pinkish marine alga strengthened with calcium carbonate deposits in the cells. North Atlantic coast from Norway to Morocco, around the UK and elsewhere.

Cordial, adj. & n. Cheering or invigorating the heart; a medicine with this property. Note: today's definition of a sweetened drink does not apply.

Coriander, Roman, n. (fennel flower, kalonji) The plant *Nigella sativa* L. [Ranunculaceae]. Mediterranean region. The "onion seeds" of naan breads.

Corm, n. Superficially similar to a bulb, but whereas a bulb is a modified bud with thickened leaves, a corm is a modified stem with a solid structure.

Costive, adj. Causing, or affected with, constipation.

Costus, n. Root of the thistle-like plant *Saussurea costus* (Falc.) Lipsch. [Compositae]. E Himalayas. Listed under CITES Appendix I.

Crab's-eye, n. White, stony, pea-sized structure (gastrolith) found in the stomach of crayfish (small fresh-water crustaceans) (and formerly thought to be from the animals' heads; James, 1747).

Crocus martis, n. Burnt iron sulphate (Mayne, 1860).

Crocus martis niger, n. (*crocus martis aperitivus*) A form of iron sulphide made by rubbing hot iron with suphur and allowing the resultant drops to fall into cold water (Parr, 1819, vol. I, p. 661).

Crocus metallorum, n. (*crocus antimonii*) A chemical preparation of antimony (Mayne, 1860).

Cubeb, n. (tailed pepper, Java pepper) Fruit of *Piper cubeba* L.f. [Piperaceae]. Java & Indonesia.

Cuckoo-pint, n. The plant *Arum maculatum* L. [Araceae]. Europe.

Cucurbit, n. A gourd-shaped vessel that forms that bottom part of historical apparatus for distillation (q.v. alembic).

Cudweed, mountain, n. (mountain everlasting) The plant *Antennaria dioica* (L.) Geartn. [Compositae] (syn. *Gnaphalium dioica*). Eurasia & N America.

Cypress-nut, n. Fruit of the tree *Cupressus sempervirens* L. [Cupressaceae]. S Europe, Libya & SW Asia.

Cystic duct, n. The duct between the gall bladder and the bile duct.

Danewort, n. (dwarf elder) The plant *Sambucus ebulus* L. [Viburnaceae]. Eurasia & Mediterranean.

Diacolocynthidos, Duchesne's, n. An oil containing colocynth (q.v.) pulp and other ingredients (*Ph. Aug.*, 1684, p. 207).

Diagrydium, n. Scammony (q.v.) prepared by baking it inside a hollowed-out quince (Hill, 1751). It could then be further treated with sulphur to make **sulphurated diagrydium** (Culpeper, 1683, p. 264).

Diaphoretic, adj. & n. Promoting sweating; a medicine with this property.

Diamoron, n. A preparation containing mulberries, *Morus nigra* L. (Culpeper, 1683, p. 131).

Diasulphur, n. A medicine containing sulphur.

Dicelon, n. An apparition sent to make men afraid (Lavaterus, 1596, Part I, p. 7).

Diploma, n. A double vessel used in chemical operations (OED).

Dipsas, n. Mythological serpent, whose bite was fabled to produce a raging thirst.

Dittany, n. (white dittany) The plant *Dictamnus albus* L. [Rutaceae]. Eurasia. Alternatively, dittany of Crete (see next).

Dittany of Crete, n. The plant *Origanum dictamnus* L. [Labiatae]. Greece & Crete.

Dodder, n. *Cuscuta* spp. [Convolvulaceae]. A cosmopolitan genus (including the UK) of parasitic flowering plants, whose reddish, thread-like stems, trail over and penetrate their host plant.

Dog's mercury, n. The plant *Mercurialis perennis* L. (Euphorbiaceae). Eurasia & Mediterraean region.

Drachm, n. Apothecaries' unit of weight, equal to one eighth of an Apothecaries' ounce, or three scruples, or 60 grains. The British Apothecaries' drachm (used until 1864) was equivalent to 3.89 g.

Dragon's blood, n. A deep red, resinous substance derived from various plants: (i) **Canaries dragon's blood**, *Dracaena draco* (L.) L. of the Canaries, Madeira, S Morocco & Cape Verde; (ii) **Socotran dragon's blood,** *Dracaena cinnabari* Balf.f. [Asparagaceae] of Socotra (the original kind of dragon's blood, Flückiger & Hanbury, 1874); or (iii) **Sumatran dragon's blood**, *Calamus macrophylla* (Becc.) W. Bak. or *C. verticillaris* Griff [Palmae] (syns. *Daemonorops macrophylla* and *D. verticillaris*) of the Malay Peninsula.

Dyspnoea, n. Shortness of breath; breathing difficulty.

Ecchymosis, n. A bruise, or the leaking of blood into surrounding tissues because of a bruise.

Eclegme, n. (eclegma) A medicine in the form of a thick syrup, to be taken by licking off a liquorice stick held in the mouth (Tomlinson, 1657).

Elecampane, n. The plant *Inula helenium* L. [Compositae]. C Asia & Europe.

Electuary, n. Variously defined as a syrup-like medicine, and/or a medicine specifically used as an antidote.

Electuary of diasatyrion, n. A preparation of satyrion roots (q.v.) and other ingredients (Culpeper, 1683, p. 162).

Electuary of diascordium, n. A preparation of water germander (q.v.) and other ingredients (Culpeper, 1683, p. 164).

Elemi, n. (gum elemi) A gum sourced from various plants of the Burseraceae family, either: (i) **African elemi** (the original kind of elemi) *Boswellia frereana* Birdw. of Somalia (Flückiger & Hanbury, 1874); (ii) **American elemi**, *Bursera simaruba* (L.) Sarg. of S Florida and C America; (iii) **West Indian elemi**, *Dacryodes excelsa* Vahl.; or alternatively (iv) *Amyris elemifera* L. [Rutaceae] of Tropical America (Mayne, 1860).

Emmenagogue adj. & n. Promoting menstruation; a medicine with this property.

Empusa, n. In Greek mythology, a female demon associated with the goddess Hectate (OED); an apparition that can change its shape and which mostly appears at noon to "such as are in misery" (Lavaterus, 1596, Part I, p. 6).

Empyema, n. Originally an accumulation of pus inside a structure, but later the definition was narrowed to mean an accumulation of pus within the thoracic cavity.

Empyreuma, n. The peculiar smell (and taste) arising from the products of decomposition of biological substances burnt in closed containers.

Endive, n. The plant *Cichorium endivia* L. [Compositae]. Mediterranean.

Ephialtes, n. Synonymous with incubus (q.v.) (Lavaterus, 1596, Part I, p. 6).

Epicureans, n. Followers of the Greek philosopher Epicurus who thought, amongst other things, that the gods did not concern themselves with human affairs.

Epithem, n. An externally-applied medicine, especially one with alterative (altering processes within the body, especially as opposed to causing excretion or evacuation) and alexiteric (q.v.) properties (Tomlinson, 1657, book V, p. 190).

Ethmoid bone, n. A bone forming part of the orbit of the eye. The **cribriform plate** of the ethmoid bone forms the roof of the nasal cavity.

Eupatorium, n. This word could signifiy different species but without a qualifying adjective (as occurs in some places in *Crocologia*) it is not possible to identify which plant is meant. The different kinds are: (i) **"eupatorium of the Ancients"** (Dioscorides, Galen etc.), namely agrimony (q.v.); (ii) **"Avicenna's eupatorium"**, namely hemp agrimony, *Eupatorium cannabinum* L. [Compositae] (Miller, 1722); or (iii) **"Mesue's eupatorium"**, namely common maudlin (or English mace) *Achillea ageratum* L. [Compositae]. S Europe. Blancard (1708) defines eupatorium as hemp agrimony solely.

Euphorbium, n. Inspissated juice of the thorny succulent *Euphorbia resinifera* O. Berg. [Euphorbiaceae]. Morocco.

Excrementitious, adj. Of or pertaining to excrement or excreted matter.

Expectorant adj. & n. Promoting the coughing up of phlegm; a medicine with this property.

Extravasated, adj. A fluid, especially blood, run or forced out of its normal vessel.

Eyebright, n. The plant *Euphrasia officinalis* L. [Orobanchaceae]. Europe.

Faecula, n. The lees. Faecula of bryony was made as follows: the juice was pressed from the roots and stored in vessels until the lees were deposited at the bottom, then the fluid was poured away and the lees dried (Pechey, 1707). Other roots were treated similarly, and essentially the same process is used today to make arrowroot.

Falernian wine, n. A wine from Campania, celebrated by the Romans.

Faunes, n. Monsters inhabiting woods, the top half like a man, the lower half like a goat. Identical to panes (q.v.) (Lavaterus, 1596, part I, pp. 6–7).

Fenugreek, n. The plant *Trigonella foenum-graecum* L. [Leguminosae]. S Europe & W Asia.

Feverfew, n. The plant *Tanacetum parthenium* (L.) Schultz-Bip. [Compositae]. Europe.

Foment, v. To cherish with heat; to bathe with warmed lotions.

Frontal, n. (frontale) A topical medicine applied to the forehead, often in a special linen bag.

Galangal, n. Dried rhizomes of either of two conspecific plants from the Zingiberaceae family, the true identity of which was not discovered until 1870 (Flückiger & Hanbury, 1874): (i) **greater galangal**, *Alpinia galanga* (L.) Willd. of tropical Asia; or (ii) **lesser galangal**, *A. officinarum* Hance of E & SE Asia. The lesser galangal was more aromatic (Miller, 1722).

Galbanum, n. A gum-resin from the Iranian umbelliferous plant species *Ferula gummosa* Boiss., *F. rubricaulis* Boiss. and *F. galbaniflua* Boiss. & Buhse (Flückiger & Hanbury, 1874). (*F. galbaniflua* is an as yet unresolved name that could be designated a synonym of *F. gummosa* in future).

Galingale, English, n. (sweet cyperus) Roots of *Cyperus longus* L. [Cyperaceae]. Eurasia.

Gallia moschata, n. The name of a compound medicine (Culpeper, 1683, p. 195).

Gentian, n. The plant *Gentiana lutea* L. [Gentianaceae]. Europe & W Asia.

Geophyte, n. A perennial plant that grows from an underground organ, e.g. a bulb, corm (q.v.) or tuber.

Germander, n. The plant *Teucrium chamaedrys* L. [Labiatae]. Europe to the Caucasas.

Germander, felty, n. (poley mountain) The plant *Teucrium polium* L. [Labiatae]. Mediterranean to SW Asia.

Germander, greater, n. Possibly *Teucrium flavum* L. (syn. *T. majus*) [Labiatae]. Europe (Parkinson, 1640).

Germander, water, n. The plant *Teucrium scordium* L. [Labiatae]. Eurasia to Mediterranean area.

Gersa serpentaria, n. A substance prepared from "aaron roots" (cuckoo-pint, q.v.) in the same way that faecula (q.v.) of bryony is made (Culpeper, 1683, p. 262).

Grain, n. Apothecaries' smallest unit of weight, equal to one twentieth of a scruple, or one sixtieth of a drachm. There are 180 grains in an Apothecaries' ounce. The British Apothecaries' grain (used until 1864) was equivalent to 0.0648 g.

Grain of paradise, n. Dried fruit of *Aframomum melegueta* (Roscoe) K. Schum (syn. *Amomum grana-paradisi*) [Zingiberaceae]. W Africa.

Grayling, n. *Thymallus thymallus* L., [Salmonidae], a freshwater fish reaching up to 0.6 m in length.

Ground-ivy, n. (alehoof) The plant *Glechoma hederacea* L. [Labiatae]. Europe.

Ground-pine, n. The plant *Ajuga chamaepitys* (L.) Schreb. of Eurasia, and possibly also *A. iva* (L.) Schreb. of the Mediterranean region [Labiatae].

Grumous, adj. Thick, clotted, viscid.

Guaiacum, n. Resin from *Guaiacum officinale* L. & *G. sanctum* L. [Zygophyllaceae]. Jamaica & the W Indies. The timber of these trees is lignum vitae.

Harpy, n. A mythical beast with a woman's face and body, and bird's wings and claws (OED); a "ravening" devil with the face of a maiden (Lavaterus, 1596, part I, p. 7).

Hartwort, n. (ser-mountain) The plants *Lasterpitium siler* L., and possibly also *Tordylium* spp. (Mabberley, 2017) and *Seseli tortuosum* L. (Mayne, 1860) [all Umbelliferae]. Europe.

Hellebore, black, n. The plant *Helleborus niger* L. [Ranunculaceae]. The Alps & Appenines.

Hellebore, white, n. The plant *Veratrum album* L. [Melanthiaceae]. Eurasia.

Henbane, n. (black henbane) The plant *Hyoscyamus niger* L. [Solanaceae]. Temperate Eurasia.

Hepatica, n. Either of two unrelated plant species: (i) *Anemone hepatica* L. [Ranunculaceae], Eurasia; or (ii) a species of liverwort, *Marchantia polymorpha* L. (listed as the *hepatica vulgaris* by Mayne (1860)).

Hermodactyl, n. A dried root from Turkey, described as light brown with little smell or taste (Miller, 1722), but a very characteristic morphology (carefully detailed by Hill, 1751). The source-plant is still uncertain; it has been suggested to be the roots of meadow saffron (*Colchicum autumnale*; James, 1747) or of species of *Iris, Cyclamen* or (more likely) a *Colchicum* species from the East (Flückiger & Hanbury, 1874).

Hiera picra, n. [Greek = "sacred bitter"] A preparation of aloes and other ingredients (Culpeper, 1683, p. 174).

Hippotaur, n. Probably Hertodt's rendering of the word "hippocentaur", a mythological creature with the upper body of a man and the body and legs of a horse.

Hippocratic sleeve, n. A long, narrow cloth bag used as a filter.

Horehound, n. (white horehound) The plant *Marrubium vulgare* L. [Labiatae]. Eurasia & Mediterranean.

Houseleek, n. The succulent plant *Sempervivum tectorum* L. [Crassulaceae]. Eurasia.

Hyacinth, n. The plant, *Hyacinthoides non-scripta* (L.) Rothm. (syn. *Scilla non-scripta*) the common bluebell [Asparagaceae] of W Europe. Alternatively a gemstone: orange-red forms of the mineral zircon. London druggists apprarently kept small garnets (also a type of neosilicate, like zircon) the size of grains of sand as "hyacinth" (Culpeper, 1683).

Hydrocephalus, n. Water on the brain, a disease especially of young children.

Hyphialtae, n. Synonymous with succubus (q.v.) (Lavaterus, 1596, part I, p. 6).

Hypochondrium, n. The upper part of the abdomen on either side. The liver in is the right hypochondrium.

Hypocistis, n. (hypocistidis) Sun-dried juice pressed from the bruised flowers of *Cytinus hypocistis* (L.) L. [Cytinaceae] (Pechey, 1707), a parasitic plant from the Mediterranean.

Hyposphagma, n. Obsolete term for a bruise or contusion of the eye, a bloodshot eye.

Hyssop, n. The plant *Hyssopus officinalis* L. [Labiatae]. Temperate Eurasia.

Incubus, n. A male spirit that sought to have intercourse with sleeping women (q.v. succubus); a feeling of oppression during sleep; a nightmare.

Induration, n. A localised hardening of soft tissue.

Infarct, n. A portion of tissue that has become stuffed with extravasated blood, serum, or other matter.

Insolation, n. Exposure of some substance to the sun.

Inspissate, v. To thicken by dehydration.

Iris, n. *Iris* spp. [Iridaceae]. A garden plant with purplish-blue flowers. The juice of the root was the only part used (Miller, 1722). Also see Orris.

Ivy, gum of, n. Gold-coloured gum collected from wounds in thick and woody ivy (*Hedera helix* L. [Araliaceae]) stems.

Jalap, n. (black mechoacan) Dried root of *Ipomoea purga* (Wender.) Hayne [Convolvulaceae]. Mexico. See also Mechoacan.

Jew's-ear fungus, n. The ear-like fruiting body of *Auricularia auricula-judae* (Bull.) J. Schröt., which often develops on elder trees (*Sambucus nigra* L. [Viburnaceae]).

Jujube, n. Date-like dried fruit of *Ziziphus jujuba* Mill. [Rhamnaceae]. SE Europe to China.

Julep, n. A sweet drink used as a vehicle for medicine.

Kermes-berry, n. Harvested from the kermes oak *Quercus coccifera* L. [Fagaceae], a Mediterranean tree (Pechey, 1707), the "berry" is a species of scale insect, *Kermes vermilio* Planchon, yielding a brilliant red dye. Cochineal is produced from a related New World scale insect species.

Lac, n. (gum lac, shellac) This substance is typically listed under plant products in early books, produced from "Jujuba Indica" (e.g. Miller, 1722, p. 248). It was not understood that lac is a product of insects, mainly *Kerria lacca* Kerr from India. Three kinds of lac were in commerce: (i) **seed-lac**, small reddish grains; (ii) **shell-lac** (shellac), the best kind, in flat, thin transparent pieces (Pechey, 1707); and (iii) **stick-lac**, where the lac is covering the surfaces of short sticks (Miller, 1722).

Ladanum, n. (labdanum) Leaf and stem resin of the shrub *Cistus ladanifer* L. [Cistaceae]. W Mediterranean region.

Larkspur, n. The plant *Consolida regalis* Gray [Ranunculaceae]. Europe & W Asia.

Laudanum, n. The name for various preparations in which opium was a major ingredient; tincture of opium.

Leopard's bane, n. Plants of the genus *Doronicum*, especially *D. pardalianches* L. [Compositae]. Europe.

Lily of the valley, n. The plant *Convallaria majalis* L. [Asparagaceae]. N temperate Europe.

Lily, n. (white lily) *Lilium candidum* L. (syn. *L. album*) [Liliaceae]. Originally from Crete, Lebanon & Israel.

Liquidambar, n. Balsam from *Liquidambar styraciflua* L. [Hamamelidaceae]. Connecticut to C America.

Litharge, n. A red-yellow oxide of lead.

Lithontriptic, adj. & n. Breaking up stone(s) in the bladder; a medicine with this property.

Lixivial, adj. Of or pertaining to lye; alkaline.

Lochia, n. (cleansings) The flow of discharge from the uterus and vagina some days after the delivery of a baby.

Lohoch, n. (looch) Sometimes used as synonym for eclegme (q.v.), but historically a lohoch was an eclegme made for instant use, rather than for storage (Tomlinson, 1657, book v, p. 165).

Lohoch sanum, n. The name of a type of lohoch (q.v.) (Zwelfer, 1672, p. 176 & 558).

Lovage, n. The plant *Levisticum officinale* Koch. [Umbelliferae]. Europe.

Lungwort, n. (Jerusalem cowslip) The plant *Pulmonaria officinalis* L. [Boraginaceae]. Europe & W Asia.

Lupin, n. (white lupin) The plant *Lupinus albus* L. [Leguminosae]. S Europe.

Lute, n. & v. A type of cement to make an air-tight joint in a retort or other piece of apparatus; the act of applying lute.

Lycium, n. A substance of uncertain identity, exported from India since Antiquity. It was thought to be a product of the wood or roots of Indian species of *Berberis* [Berberidaceae] (Flückiger & Hanbury, 1874), or from the box-thorn, *Lycium barbarum* L. [Solanaceae].

Mace, n. Spice comprising the fimbriate aril surrounding a nutmeg.

Madder, n. The plant *Rubia tinctorum* L. [Rubiaceae]. Roots were mainly used as a red dyestuff, with a more limited application in medicine (Miller, 1722).

Magistery, n. General term for a white powder prepared from animal, vegetable or mineral starting ingredients; the concentrated essence of a substance.

Magistral, adj. & n. Applied to medicines prescribed for a particular occasion, as apposed to officinal medicines which were pre-preated and kept in shops (Mayne, 1860). The OED further adds: a medicine considered to be supremely effective, or a preparation devised by a physician for a particular case (not included in an official pharmacopoeia).

Maidenhair, n. The cosmopolitan fern species, *Adiantum capillis-veneris* L. [Pteridaceae].

Mallow, n. The plant *Malva sylvestris* L. [Malvaceae]. Europe.

Mandrake, n. The plant *Mandragora officinarum* L. [Solanaceae]. S Europe.

Mandrake-apple, n. (*pomis mandragorae*) The fruit of the mandrake plant (Tomlinson, 1657, p. 656) (see above).

Manichean, n. Follower of Manichaeism, a religion common in the Roman Empire and Asia until the 5th Century, based on the duality of light and dark, good and evil.

Manna, n. Product of the tree *Fraxinus ornus* L. [Oleaceae]. S Europe & W Asia. The best manna came from Calabria (Blancard, 1708).

Manus Christi, n. A mixture of sugar and rose-water (*Ph. Aug.*, 1684, p. 142); a kind of sugar (Blancard, 1708).

Marigold, n. (pot marigold) The plant *Calendula officinalis* L. [Compositae]. Mediterranean region.

Marjoram, wild, n. The plant *Origanum vulgare* L. [Labiatae]. Europe to C Asia.

Marocostinum, Minderer's, n. A purgative extract of *Teucrium marum* L. (marum germander) and costus (q.v.) (Mayne, 1860; Culpeper, 1683, p. 182).

Marshmallow, n. The plant *Althaea officinalis* L. [Malvaceae]. Europe.

Mastic, n. Gum from the plant *Pistachia lentiscus* L. [Anacardiaceae], traditionally harvested only on the Greek island of Chios (Miller, 1722).

Maturative, adj. & n. Promoting the formation of pus (suppuration); a medicine with this property.

Mechoacan, n. (white jalap) Dried root of *Ipomea purga* (Wender.) Hayne [Convolvulaceae] (syn. *I. mechoacan*), similar to jalap (Miller, 1722) (q.v.).

Melitean earth, n. A type of earth sourced from the island of Melitea (Tomlinson, 1657, book II, p. 397), now called Mljet (one of the Adriatic Islands).

Melilot, n. The plant *Melilotus officinalis* (L.) Pallas [Leguminosae]. Eurasia.

Menses, n. The menstrual discharge.

Menstruum, n. Now taken to mean a solvent, but historically also a liquor to be distilled (Blancard, 1708).

Mercurial, n. A preparation of mercury.

Mercury, dulcified, n. (calomel) Hg_2Cl_2 A tasteless, purgative, preparation of mercury made from corrosive sublimate of mercury "dulcified" by the addition of crude mercury. The most frequently used mercury preparation in medicine according to Motherby & Wallis (1745).

Miasma, n. A kind of invisible vapour or mist emanating from marshes, rotting organic material etc. causing diseases. Miasma theory remained current in Western Medicine until displaced by Germ Theory after the work of John Snow, Louis Pasteur and others in the nineteenth century.

Mill-dust, n. The very finest flour dust that settled over the surfaces inside a mill (Mosan, 1605).

Mithridate, n. (confection of Damocrates) An expensive, compound medicine comprising a large number of ingredients, named after King Mithridates of Pontus and Bithynia, who experimented with the properties of simples separately, and then in combination.

Mixtura simplex, n. The name of a certain medicine (Salmon, 1694, p. 46).

Mock sun, n. (sundog or parhelion) A bright spot on either or both sides of the sun, caused by light interacting with ice crystals in the atmosphere.

Mola, n. (mole) An abnormal mass in the uterus; a false conception.

Moravian measure, n. A unit of volume. Gyllenbok (2018, p. 1764) gives various units of liquid capacity used in Brno (in Moravia: where Hertodt was living when he wrote *Crocologia*) in the sixteenth and seventeenth centuries. Those that were probably meant on p. 60 ("13 Moravian measures" of spring water) were either the *mass* (1.0896 litres equivalent) or the *seitl* (272.322 ml).

Moschardinus, n. A troche (q.v.) containing musk (q.v.).

Mouse, mountain, n. An animal from the high Alps, probably a marmot, whose fat was "recommended in nervous Affections, and for Stiffness and Contractions of the Joints" (James, 1747). Probably "la marmotte des Alpes", *Marmota marmota* L. and possibly also "la marmotte de Pologne" (Brisson, 1756, p. 165), *M. bobak* Müller (syn. *M. polonica*).

Mouse-ear hawkweed, n. The plant *Pilosella officinarum* Vaill. [Compositae]. Europe & Mediterranean.

Mugwort, n. The plant *Artemisia vulgaris* L. [Compositae]. N temperate regions.

Mullein, n. (white or great mullein, Aaron's rod) The plant *Verbascum thapsus* L. [Scrophulariaceae]. Eurasia.

Mummy, n. True mummy was a fragrant liquid, with the consistency of honey, that was found in the coffins of mummified Egyptian kings and noblemen (Tomlinson, 1657, book III, pp. 441–442). Demand exceeded supply and so fake mummy was made from "the flesh of dried bodies", or from the "succe [juice] expressed out of putrid carcasses" (Ibid.).

Mundificative, adj. & n. Cleansing impurities from wounds or sores; a medicine with this property.

Myrobalan, chebulic, n. Dried fruits of *Terminalia chebula* Retz. [Combretaceae]. Indomalasia.

Myrtillus, n. Either: (i) **myrtle** (Mosan, 1605), *Myrtus communis* L. [Myrtaceae], of the Mediterranean region; or (ii) **bilberries** (Miller, 1722), *Vaccinium myrtillus* L. [Ericaceae] of Eurasia.

Nard, n. (spikenard, Indian nard) The root of *Nardostachys jatamansi* (D.Don) DC. (syn. *N. grandifolia*) [Caprifoliaceae]. Himalayas.

Nard, Celtic, n. The plant *Valeriana celtica* L. [Caprifoliaceae]. Alps.

Nard, Indian, n. see Nard.

Nepenthes, n. A drug to "drive away grief or sadness" comprising many ingredients including saffron and opium (Mayne, 1860; *nepenthes opiatum* in Culpeper, 1683, p. 187).

Nereid, n. A sea monster with the face of a man (Lavaterus, 1596, Part I, p. 7); a sea nymph (OED).

Nightshade, n. The plant *Solanum nigrum* L. [Solanaceae]. Eurasia.

Nodulus, n. A medicine enclosed in a knot in a rag, with the purpose of suspending the nodulus in a liquid for infusion (Parr, 1809).

Nubecula, n. A small corneal opacity.

Nut, Indian, n. Possibly the areca, or betel, nut (James, 1747), *Areca catechu* L. [Palmae]. The term "Indian nut" was also used historically for nutmegs and coconuts (Flückiger & Hanbury, 1874).

Nux vomica, n. (obsolete name: vomic nut) Seeds of *Strychnos nux-vomica* L. [Strychnaceae]. S Asia. Source of the poison strychnine.

Odontalgia, n. Toothache.

Oloesaccharum, n. A mixture of essential oil and sugar.

Onocentaur, n. Mythological creature with the upper body of a man and the body and legs of an ass.

Onosceli, n. Probably meaning "onoskelis", a mythological creature with one leg of an ass, the other made of brass (Smedley et al., 1855, p. 72).

Opopanax, n. (opoponax) Product of the plant *Opopanax chironium* Koch. [Umbelliferae]. Syria.

Orbicule, n. A small ball.

Organoleptic, adj. Of or relating to smell and taste.

Orpiment, n. The toxic, yellow-orange mineral, arsenic sulphide, As_2S_3.

Orris, n. Dried rhizome of *Iris* x *germanica* L. Florentina group (syn. *I. florentina*) [Iridaceae]. See also Iris.

Ounce, n. Here meaning an Apothecaries' ounce, of which 12 made an Apothecaries' pound. The precise weight of the Apothecaries' ounce varied over Europe (Gyllenbok, 2018). The British Apothecaries' ounce (used until 1864) was equivalent to 31.1 g.

Oxymel simplex, n. A mixture of honey and vinegar (Mayne, 1860; *Ph. Aug.*, 1684, p. 31).

Palla marina, n. (sea ball) Spherical mass of dead material formed by being rolled on beaches by the waves. Derived from *Posidonia oceanica* (L.) Del. [Posidoniaceae], a marine plant of the Mediterranean.

Panchymagogum, n. A purgative extract, said to purge away all kinds of humours (Blancard, 1708).

Pandaleon, n. The name of a solid paste medicine, often prescribed for lung problems (Tomlinson, 1657, book v, p. 172).

Panes, n. Monsters inhabiting woods, the top half like a man, the lower half like a goat. Identical to Fauni (q.v. faunes) (Lavaterus, 1596, part I, pp. 6–7).

Panic, n. Miller (1722) describes this as a grain crop widely sown in Germany, with seeds smaller than common millet. Probably *Setaria italica* (L.) P. Beauv. [Gramineae] (syns *Panicum germanicum*, *P. italicum*), foxtail millet.

Pannus, n. Abnormal flap of tissue, especially covering the cornea of the eye.

Paroxysm, n. A fit of a disease; a sudden increase in the severity of symptoms.

Pecten, n. (pectin pubis) The pubic bone.

Pectineal bone, n. A particular part of the pelvis.

Pellitory of the wall, n. The plant *Parietaria judiaca* L. [Urticaceae] (syn. *P. diffusa*). W & S Europe.

Pellitory of Spain, n. The plant *Anacyclus pyrethrum* (L.) Lag. [Compositae]. Mediterranean region.

Penide, n. A type of candy made of sugar and barley-water, and used against coughs and colds (Tomlinson, 1657, Book III, p. 614); a kind of clarified sugar (Blancard, 1708).

Pennyroyal, n. The plant *Mentha pulegium* L. [Labiatae]. W Europe & Asia.

Peony, n. The plant *Paeonia officinalis* L. [Paeoniaceae]. Europe.

Pepper, long, n. Seeds of *Piper longum* L. [Piperaceae]. E Himalayas.

Pepper, water, n. (obsolete name: arsesmart) The plant *Persicaria hydropiper* (L.) Delarb. [Polygonaceae]. Widespread globally.

Perch stone n. Walton (1653, p. 179) writes of perches (the freshwater fish, *Perca fluviatilis* L.), "... they have in their brain a stone, which is in forrain [foreign] parts sold by Apothecaries, being there noted to be very medicinable against the stone on the reins [kidney stone]".

Petechial, adj. Pertaining to petechiae, small spots caused by subcutaneous bleeding.

Pharmacokinetic, adj. Of or relating to the processes of absorption, distribution, metabolism and excretion of a drug.

Philonium, Roman, n. A narcotic antidote (Mayne, 1860; Culpeper, 1683, p. 165).

Phlegmagogue, n. A medicine for expelling phlegm.

Phlycten, n. Earlier definition: small watery blisters anywhere on the body, or a small pustule of the conjunctiva; later, a blister of the eye.

Phthisis, n. A severe wasting of the body, often the result of pulmonary tuberculosis.

Pica, n. An appetite for non-food items e.g. soil, hair.

Pill-mass, n. A mixture of ingredients before it is divided up and made into pills.

Pilula, n. A pill. See also orbicule.

Pipkin, n. (*ollula* in original version of *Crocologia*) An earthenware cooking vessel with three feet and a handle (Entinck, 1771).

Pituitous, adj. Of of pertaining to phlegm; phlegmatic.

Polychrestic, adj. A medicine used to treat different diseases; a medicine with many virtues.

Polypodium (of oak), n. (common polypody) The fern *Polypodium vulgare* L. [Polypodiaceae]. Eurasia. That growing on oak trees was considered best (Miller, 1722).

Pomade, n. An aromatic ointment.

Poppy, white, n. White-flowered forms of the plant *Papaver somniferum* L. [Papaveraceae] (species also has red, pink etc. flowers).

Poppy, wild, n. (corn poppy) *Papaver rhoeas* L. [Papaveraceae]. Eurasia.

Procidentia, n. A prolapse.

Prunella, n. Obsolete name for a tongue or mouth disease associated with a brown membranous coating developing on the tongue. The words seems to have disappeared from English books in the eighteenth century.

Psyllium, n. (fleawort) Seeds of *Plantago afra* L. (syn. *P. psyllium*) [Plantaginaceae]. Mediterranean to India.

Pterygium, n. An abnormal triangular growth of fibrovascular tissue over part of the eye.

Pugil, n. A unit of volumetric measurement defined as an eighth of a handful.

Quartan, n. A fever with symptoms that recur every four days.

Quartarius, n. A unit of liquid capacity originally equal to a quarter of a sextarius (q.v.).

Quince, n. Fruit of *Cydonia oblonga* Mill. [Rosaceae]. Europe and SW Asia.

Quotidian, n. A fever with symptoms that recur every day.

Racemose, adj. In the form of a cluster; resembling a bunch of grapes.

Recrement, n. A waste product.

Rheum, n. A thin, watery secretion from the eyes, nose or mouth. Earlier thought to come directly from the brain.

Rhubarb, n. Not the garden rhubarb, but the related, larger, medicinal species *Rheum officinarum* Baill. [Polygonaceae]. W China & Tibet.

Rob, n. Inspissated juice (often from fruit) evaporated to the consistency of honey or oil.

Rue, n. The plant *Ruta graveolens* L. [Rutaceae]. S Europe.

Safflower, n. the plant *Carthamus tinctorius* L. [Compositae]. The Near East.

Sagapenum, n. (gum sagapenum) A gum-resin historically imported from Alexandria (Miller, 1722; James, 1747). Virtually absent from the market by the Victorian period, but thought to have been produced by *Ferula* spp., possibly *F. persica* Willd. [Umbelliferae] (Mayne, 1860; Mabberley, 2017).

Sal martis subdulcis, n. Substance made by dissolving sal martis (green vitriol, ferrous sulphate) in spirit of wine, then burning off the spirit (Mojon, 1783, p. 142).

Sal-ammoniac, n. Ammonium chloride, originally collected from springs in Libya. Not to be confused with gum ammoniac or ammoniacum (q.v.).

Sal-prunella, n. Sustance made from nitre and sulphur (Motherby & Wallis, 1745).

Sanders, n. (saunders) Three kinds of sanders woods are specified in *Crocologia*: (i) **red sanders**, the unscented (Miller, 1722; Flückiger & Hanbury, 1874) red timber of *Pterocarpus santalinus* L.f. [Leguminosae] from Coromandel & the E Indies; (ii) **white sanders**, scented sapwood of the sandalwood tree, *Santalum album* L. [Santalaceae] of Timor and the E Indies; or (iii) **yellow sanders**, the scented heartwood of *S. album*.

Sanicle, n. The plant *Sanicula europaea* L. [Umbelliferae]. Europe.

Sape, n. (sapa) A syrup made of the juice of unripe grapes (Tomlinson, 1657, p. 533). Sometimes used as a synonym for rob (q.v.).

Sarcocolla, n. [Greek = "flesh glue"] Gum from *Astragalus gummifer* Lab. [Leguminosae]. It was normally prepared by being "nourished" in milk prior to use (Tomlinson, 1657, book II, p. 58).

Sarsaparilla, n. Roots of *Smilax* spp. including *S. officinalis* Kunth [Smilaceae]. NC America.

Sassafras, n. Root of *Sassafras albidum* (Nutt.) Nees. [Lauraceae]. EN America.

Satyr, n. In Greek or Roman mythology, beings living in woodland, and which are part man, part animal (OED); Lavaterus (1596, Part I, p. 7) implies satyrs are more human-like than e.g. panes (q.v.).

Satyrion, n. Tubers of several species of orchids [Orchidaceae] from Europe and W Asia (fully listed by Flückiger & Hanbury, 1874).

Savin, n. (savine) The plant *Juniperus sabina* L. [Cupressaceae]. C & S Europe to W Asia. "Grains of savin" mentioned on p. 193 may refer to the fruits, which are pea-sized and covered with a bluish waxy bloom, and have an essential oil content approximately four times that of the more usually used leaves (Flückiger & Hanbury, 1874).

Savoury, n. The plants *Satureja hortensis* L. (summer savoury) or *S. montana* L. (winter savoury) [Labiatae]. Mediterranean region & S Europe.

Scabious, n. The plant *Knautia arvensis* (L.) Coulter [Caprifoliaceae]. Europe to N Africa.

Scammony, n. Dried root sap (Flückiger & Hanbury, 1874) of *Convolvulus scammonia* L. [Convolvulaceae]. SW Asia.

Scelotyrbic, adj. Affecting a particular symptom of survy (Blancard, 1708, p. 262, defines "scelotyrbe" as a wandering pain in the legs caused by scurvy).

Schirrus, n. A hard tumour or swelling; later use applied to cancer.

Schoenanth, n. (squinanth) The grass *Cymbopogon schoenanthus* (L.) Spreng. (syn. *Andropogon schoenanthus*) [Gramineae]. N Africa to N India.

Scorzonera, n. (obsolete name: vipers' grass) The plant *Scorzonera hispanica* L. [Compositae]. Europe.

Scruple, n. An Apothecaries' unit of measurement equal to 20 grains, or one third of a drachm, or one twenty-fourth of an Apothecaries' ounce. The British Apothecaries' scruple (used until 1864) was equivalent to 1.296 g.

Scurvy-grass, n. The plant *Cochlearia officinalis* L. [Cruciferae]. Europe.

Scylla, n. A sea monster (OED); a "ravening" devil with the face of a maiden (Lavaterus, 1596, part I, p. 7).

Secundines, n. The afterbirth: placenta and foetal membranes.

Self-heal, n. The plant *Prunella vulgaris* L. [Labiatae]. Eurasia.

Sella turcica, n. A depression in the sphenoid bone inside the skull, associated with the pituitary gland.

Sensorium, n. The supposed centre in the brain in which sensations are united (OED).

Serosity, n. & adj. Watery humours. The condition of being serous (q.v.).

Serous, adj. Of or pertaining to serum; moistened with serum.

Serpiginous, adj. Slowly creeping, spreading.

Sextarius, n. A unit of liquid capacity, roughly equivatent to one pint, originally equal to one-sixth of a congius (OED); a sixth of a gallon (Blancard, 1708).

Sief, n. A medicine for the eyes. Later it became seen as synonymous with collyrium (q.v.), but Tomlinson (1657, book V, p. 195) classifies a sief as a dry collyrium.

Sief album, n. The name of a particular eye medicine made with or without opium (Zwelfer, 1672, p. 224).

Silenus, n. A satyr (q.v.) "of full age" (Lavaterus, 1596, part I, p. 7). Silenus was also the foster-father of Bacchus and leader of the satyrs (OED).

Skink, n. Small, lizard-like reptiles in the Scincidae family. There are several species in mainland Europe, but none native to the UK. Hill (1751) writes that the skink of the apothecaries, then often also called the sea-skink or earth-crocodile, was sourced from the banks of the River Nile and imported eviscerated and dried.

Smallage, n. Wild forms of celery, *Apium graveolens* L. [Umbelliferae].

Smaragdine, adj. Of or pertaining to emerald. Smaragdine treacle (*theriaca smaragdina*), see *Lexicon Pharmaceuticum* (1694), p. 263.

Snake-root, n. (Virginian snakeroot, serpentary) The plant *Aristolochia serpentaria* L. [Aristolochiaceae]. EN America. Another American species, black snakeroot, *Actaea racemosa* L. [Ranunculaceae] was only made known in Britain from 1696 (Flückiger & Hanbury, 1874), making it probably too late an introduction to have been included in *Crocologia*'s formulae.

Sorrel, n. The plant *Rumex acetosa* L. [Polygonaceae]. Eurasia.

Sorrel, Spanish round-leaved, n. Probably a form of French sorrel, *Rumex scutatus* L. [Polygonaceae]. Eurasia.

Southernwood, n. (lad's love, old man) The plant *Artemisia abrotanum* L. [Compositae]. E Turkey & Mediterranean area.

Sow-bread, n. (cyclamen) The root of *Cyclamen hederifolium* L. [Primulaceae]. Europe.

Spagyric, adj. Prepared according to alchemical methods.

Spanish fly, n. (cantharis, blister beetle) An iridescent green beetle, *Lytta vesicatoria* L. [family Meloidae], mainly of S Europe. It contains the potent blistering agent, cantharidin.

Species, n. A compound medicine in the form of a finely ground powder (Tomlinson, 1657).

Species aromatica rosata, n. A species containing red roses and other ingredients (Tomlinson, 1657, book III, p. 603; *Ph. Aug.*, 1684, p. 134).

Species cordiales temperata, n. A species containing aloes-wood and other ingredients (*Ph. Aug.*, 1684, p. 136).

Species de gemmis frigidis, n. A cooling species, containing gemstones (*species electuarii de gemmis frigidi* see Culpeper, 1683, p. 156).

Species diaireos, n. A species containing orris and other ingredients (Culpeper, 1683, p. 147).

Species diambra, n. A species containing ambergris and other ingredients (Culpeper, 1683, p. 148).

Species diamoschi dulci, n. A species containing musk and other ingredients (Culpeper, 1683, p. 148).

Species diatragacanthi frigidi, n. A species with cooling properties, the main ingredient of which is tragacanth (Culpeper, 1683, p. 150).

Species diatrion-santalon, n. A species containing the three kinds of sanders and other ingredients (Culpeper, 1683, p. 150).

Species liberans, n. (plural species liberantes) A species containing tormentil root and other ingredients (*Ph. Aug.*, 1684, p. 138).

Spectrophotometric, adj. Of or relating to spectrophotometry, a technique for the quantitative analysis of substances in which the absorption of light at a particular wavelength through a liquid gives a measure of the amount of a certain molecule in that liquid.

Spermaceti, n. Yellow oil from the 'spermaceti organ' in the head of sperm whales, *Physeter macrocephalus* L.

Sphacelus, n. A mass of mortified tissue (OED); distinguished from a gangrene "by the total Corruption and stink, it being also insensible to Knife and Fire" (Blancard, 1708).

Sphinx, n. A beast either in Greek mythology with the head of a woman and the winged body of a lion, or in Egyptian mythology with a male head and no wings.

Spike, n. (spike lavender) The plant *Lavandula latifolia* Medik. (syn. *L. spica*) [Labiatae]. Portugal to S France. A less hardy species than common lavender, with a poorer quality essential oil (Flückiger & Hanbury, 1874).

Spikenard see Nard.

Spirit, n. A substance made by distillation, either with or without prior fermentation. The fermentation step involved steeping the ingredient(s) in water and sugar, leaving for a time until a vinous scent was present then distilling. When spirits were made without fermentation, the ingredient was steeped in spirit of wine (Salmon, 1694).

Spleenwort, n. Generic name for ferns of the genus *Asplenium*. Medicinal species included the hart's-tongue fern, *Asplenium scolopendrium* L. of Europe, and *A. ceterach* L. (syn. *Ceterach officinarum*).

Spodium, n. Burnt ivory (James, 1747), or a substance formerly collected from copper furnaces (Hill, 1751).

Spurge, n. Plants of the genus *Euphorbia* [Euphorbiaceae] including *E. esula* L. (syn. *Esula major*) (Mayne, 1860), *E. palustris* L. (Miller, 1722) and *E. cyparissias* L. (syn. *Esula minor*) (Mayne, 1860).

Squill, n. The plant *Drimia maritima* (L.) Stearn [Asparagaceae]. Mediterranean coastal areas.

Storax, n. (styrax) A resin. If only "storax" was specified in a recipe, this was calamite storax (Pechey, 1707). That, and some other kinds of storax, occur in *Crocologia*: (i) **calamite storax,** *Styrax officinalis* L. [Styracaceae], Mediterranean region; (ii) **liquid storax,** *Liquidambar orientalis* Mill. [Hamamelidaceae], SW Asia Minor; or (iii) **red storax**, considered a different substance to the other kinds and came from India (Pechey, 1707). Hooper (1845) and Mayne (1860) write that red storax was the special term for storax in tears (naturally dripped from the tree) as opposed to lumps.

Strawberry tree, n. The plant *Arbutus unedo* L. [Ericaeae]. S Europe & Ireland.

Struma, n. A goitre; scrofula.

Succubus, n. A female spirit that sought to have intercourse with sleeping men (q.v. incubus).

Sudorific, adj. & n. Promoting or causing perspiration; a medicine with this property.

Sugar of lead, n. Lead (II) acetate, $Pb(CH_3COO)_2$.

Sumach, n. The plant *Rhus coriaria* L. [Anacardiaceae]. S Europe.

Sundew, n. The carnivorous plant *Drosera rotundifolia* L. [Droseraceae]. Europe.

Sweet flag, n. (calamus) The plant *Acorus calamus* L. [Acoraceae]. N temperate regions to India.

Sybarite, n. Inhabitant of an Ancient Greek city in southern Italy, Sybaris. These people were believed to be effeminate and obsessed with luxury.

Syncope n. A sudden reduction in the strength of the pulse and of respiration (Motherby & Wallis, 1745); failure of the heart's action resulting in loss of consciousness and sometimes death (OED).

Synovia, n. Today the word synovia refers to the joint fluid, but historically it meant any egg-white-like body fluid.

Tabulae rosarae, n. Tablets containing powdered roses, sugar and gum. See Salmon (1694) p. 866.

Tacamahac, n. (tacamahaca) A tree resin from various sources: (i) first brought over from America by Spanish explorers from a poplar-like tree (Hill 1751), probably *Populus balsamifera* L. [Salicaceae] of N America; (ii) later from the Spanish West Indies, probably from *Bursera* sp. [Burseraceae]; and (iii) afterwards discovered in "Madagascar" (Hill, 1751), maybe from *Calophyllum tacamahaca* Willd. [Calophyllaceae] of Mauritius & Reunion.

Tamarisk, n. The plants *Tamarix* spp. [Tamaricaceae]. Spain & S France.

Tansy, n. The plant *Tanacetum vulgare* L. [Compositae]. Eurasia.

Tepal, n. One of the outer parts of a flower, a collective term for when the petals cannot be differentiated from the sepals (often smaller than petals and not as brightly coloured; sepals protect the flower when in bud).

Terra sigillata, n. (seal'd earth) Little cakes of compacted medicinal clay/earth stamped with a seal, sourced from various locations. Descriptions of some different seal designs occur in Parkinson (1640) and Hill (1751). See also Bole.

Thebaic, adj. Of Thebes. The best opium came from there (Blancard, 1708).

Thistle, blessed, n. The plant *Centaurea benedicta* (L.) L. (syn. *Cnicus benedictus*) [Compositae]. Mediterranean region.

Thistle, carline, n. The plant *Carlina vulgaris* L. [Compositae]. Europe.

Thyme, wild, n. (mother-of-thyme) The plant *Thymus praecox* Opiz (syn. *T. serpyllum*) [Labiatae]. Europe & N Africa. A close relative of common thyme (*T. vulgaris*) but has a much wider geographical distribution, and is more hardy.

Toad-flax, n. The plant *Linaria vulgaris* Mill. [Plantaginaceae]. Eurasia.

Tophus, n. Concretion in the joint, caused by gout.

Tormentil, n. The plant *Potentilla erecta* (L.) Räusch. (syn. *P. tormentilla*) [Rosaceae]. Eurasia.

Tragacanth, n. (gum tragacanth) Gum of *Astragalus* spp. [Leguminosae].

Tragea, n. (tragema) A coarse powder of different ingredients used in a fumigation, or placed into a bag for infusion into a liquid; or a powder prepared with sugar (Blancard, 1708, p. 294).

Treacle, n. (theriac) Not today's culinary ingredient consisting of uncrystallized sugar syrups, but historically a highly complex pharmacological mixture whose composition varied depending on where it was made. Venice treacle (the treacle of Andromachus) contained amongst many other things, vipers.

Triton, n. A sea monster with the face of a man (Lavaterus, 1596, part I, p. 7); a merman (Smedley et al., 1855, p. 76); one of a race of inferior sea gods (OED).

Tritorium, n. Obsolete chemistry term meaning a vessel for separating liquids of different densities, e.g. water and oil (OED); a funnel (Blancard, 1708).

Troche, n. (trochisk) Little tablet or cake made by mixing medicine with sugar and gum tragacanth, then drying. Troches were thought suitable for administering medicine to children, and "to those who object to anything that is not a sweet-meat, and a preparation from the confectioner" (Motherby & Wallis, 1745).

Troches of alhandal, n. The name of a type of troches whose major ingredient was colocynth.

Tryphera magna, n. (triphera magna) The name of a particular compound medicine (Culpeper, 1683, p. 175).

Tuff, n. A soft rock derived from compacted volcanic ash.

Tunica adnata, n. Conjunctiva of the eye.

Turpentine, n. The product of various trees: (i) **common turpentine** from *Pinus sylvestris* L. [Pinaceae] resin (Miller, 1722); (ii) **cypress turpentine**, an inferior (Miller 1722) product of *Pistachia terebinthus* L. [Anacardiaceae] from the Mediterranean region (the finest turpentine from this species was Chios turpentine, made on the Greek island of that name (Miller, 1722) but later writers, e.g. Mayne (1860) did not distinguish any differences in quality between Chian and Cyprian turpentine); or (iii) **Venice turpentine** from *Larix decidua* Mill. [Pinaceae]. Alps, Carpathians and Pyrenees.

Turpeth, n. (turbith) Root of *Operculina turpethum* (L.) Silva Manso [Convolvulaceae]. India.

Tutty, n. A substance collected from inside furnaces where brass was made from copper and calamine (a zinc ore) (Hill, 1751).

Tympanites, n. Swelling of the abdomen caused by gas accumulating in the digestive tract or peritoneal space.

Ungula, n. Obsolete synonym for a pterygium (q.v.) (OED), but according to Motherby & Wallis & Wallis (1745), a fingernail-shaped accumulation of matter in the pupil of the eye (also called unguis).

Unicorn's horn, n. A substance of unknown identity to the ancients, but later recognised as "the tooth of a fish of Greenland" (Blancard, 1708), but also speculated to be a kind of spar from Germany or a fossil (Motherby & Wallis, 1745). The true identity seems to be the tusk of the narwhal (*Monodon monoceros* L.) of Arctic waters.

Valerian, n. The plant *Valeriana officinalis* L. [Caprifoliaceae]. Eurasia.

Verjuice, n. Acidic juice pressed from crab apples or unripe grapes.

Vervain, n. The plant *Verbena officinalis* L. [Verbenaceae]. Eurasia.

Vetch, n. The plant *Vicia sativa L.* [Leguminosae]. Not used as an ingredient, but mentioned in one recipe as a reference to the size that pills should be made.

Vine, wild, n. The flowers of wild vines (*lambrusci*: *Vitis vinifera* L.) were a known astringent, and occur in the recipe for Rhazes's troches (Schenck, 1607, p. 365). This ingredient is mentioned only once in *Crocologia* and then without specifying its flowers.

Viper, n. Venomous snakes in the Viperidae family. Several species are endemic to mainland Europe, including the adder *Vipera berus* L. Tomlinson (1657, book III, p. 467) writes that the heads and tails must be cut from vipers before use, and the "internals, *Spina Dorsi,* and belly must likewise be abjected." Hill (1751) recommends that vipers for medicinal use are best taken in July and August. They were used either freshly killed or were dried.

Vitriol, n. Sulphates of iron, copper or zinc (Mayne, 1860).

Vitriolated tartar, n. Potassium sulphate.

Vomit, n. As well as the everyday definition, this also refers to a substance given to induce vomiting, an emetic.

Wallflower, n. The plant *Erysimum cheiri* (L.) Crantz. (syn. *Cheiranthus cheiri*) [Cruciferae]. Europe.

Water lily, white, n. The plant *Nymphaea alba* L. [Nymphaeaceae]. Eurasia & Mediterranean.

Water lily, yellow, n. The plant *Nuphar luteum* (L.) Sm. [Nymphaeaceae]. Europe & N America.

Wormwood, n. The plant *Artemisia absinthum* L. [Compositae]. Eurasia & N Africa.

Wormwood, Roman, n. (pontic wormwood) The plant *Artemisia pontica* L. [Compositae]. SE Europe.

Xenexdo, n. Hertodt's rendering of the word "zenechdon", a preparation containing saffron and arsenic amongst other ingredients (Grüling, 1665, p. 521).

Yarrow, n. The plant *Achillea millefolium* L. [Compositae]. N temperate regions. The red-flowered form of this herb is specified once in *Crocologia.*

Zedoary, n. (white turmeric, set-wall) Rhizome of the plant *Curcuma zedoaria* (Christm.) Roscoe [Zingiberaceae]. NE India.

Crocologia, Our References

Abl, Dr. (1854) Ueber *Crocus sativus. Oesterreichische Zeitschrift für Pharmacie*. pp. 89–95.

Aelian (1958) *On the characteristics of animals, with an English translation by A.F. Scofield*. Harvard University Press.

Agayev, Y.M.-O., Fernández, J.A. & Zarifi, E. (2009) Clonal selection of saffron (*Crocus sativus* L.): The first optimistic experimental results. *Euphytica* 169(1): 81–99.

Ait-Oubahou, A. & El-Otmani, M. (1999) Saffron cultivation in Morocco. In: *Saffron,* Crocus sativus *L*. Ed. M. Negbi. Harwood Academic Publishers, Amsterdam. pp. 87–94.

Alexander, W. (1768) *Experimental essays*. London.

Aldrovandi, U. (1602) *De animalibus insectis librum septem*. Bologna.

Alonso Díaz-Marta, G.L., Arghittu, A., Astraka, K., Betza, T., Camba, E., Caňadas Sanchez, W., Carmona Delgado, M., Cilloco, M.T., Corona, J., Curreli, M., Daferera, D., Dejuan Valero, A., Fallas, Y., Fenza, G., Floris, I., Ibba, G., Ibba, J., García López de Rodas, E., Garrido Garcia, J., Gkoutzios, V., Kanakis, C., Laconi, R., Leoudis, A., López Córcoles, H., López Fuster, P., Lusso, G., Manconi, M., Marras, W., Mulas, G., Murgia, G., Onnis, M.V., Peddis, R., Pes, I., Picornell Buendia, R., Pinna, M.E., Pirastu, A., Pirisi, A.E., Polissiou, M.G., Podda, S., Rakitzis, K., Remoundos, P., Romero Del Rey, M.J., Sajardo Lucas, T., Sanchez Gomez, A.M., Salinas Fernandez, R., Sanna, F., Sanna, S., Satta, B., Selis, T., Tarantilis, P.A., Venusti, M., Zalacain Aramburu, A., Amor Chico, J.I., Gkola, R., Kaliva, E., Ortega Ripoll, R., Patsiouras, N., Rodríguez Rodríguez, E., Sirera Martínez, S., Valero Ganau, M., Fiori, M., Paschino, F. & Pitzalis, P. (2006) *White Book. Saffron in Europe: Problems and strategies for improving the quality and strengthen competitiveness*. [Known amongst saffron researchers as simply "The White Book"]. Available at http://www.europeansaffron.eu/archivos/White%20book%20english.pdf

Alsayied, N.F., Fernández, J.-A., Schwarzacher, T. & Heslop-Harrison, J.S. (2015) Diversity and relationships of *Crocus sativus* and its relatives analysed by inter-retroelement amplified polymorphism (IRAP). *Annals of Botany* 116(3): 359–368.

Anon. (1676a) De correspondentiâ in Holsatiam, Lusatiam & Misniam. *Miscellanea curiosa medico-physica* 4 & 5: Chapter XXXI, p. 56.

Anon. (1676b) De correspondentiâ in Moraviam, Palatinatum & Prussiam. *Miscellanea curiosa medico-physica* 4 & 5: Chapter XXXII, p. 57.

Anon. (1788) Article X. Pharmacopoeia Collegii Regalis Medicorum Londinensis. *The Analytical Review* II: 190–202.

Anon. (1876) Report of a letter by Mr Samuel Elliott read at the pharmaceutical meeting of 6th December 1876. *Pharmaceutical Journal & Transactions* VII: 482.

Anon. (1933) BMJ Epitome of current medical literature no. 273. *British Medical Journal* 1933 (2): E 63.

Anon. (1935) Import duties on drugs. *British Medical Journal* 1935 (1): 476.

Ardensbach von Ardensdorf, Wenzel Maximilian (1671) *Animadversaria in Crocologiam Hertodtianam*. In: *Tartaro-Clypaeus, excipiens Tartaro-Mastigem Hertodtianum, cui accedit discursus in Opus mirificium VI. diei, item Animadversatio in Crocologia Hertodtianam*. Prague.

Aykurt, C., Deniz, İ.G., Başaran, M.A. & Genç, G.E. (2015) Taxonomic and GIS-based ecologic studies on critically endangered *Crocus wattiorum* B. Mathew (Olympos saffron). *Thaiszia Journal of Botany* 25(1): 73–84.

Bagur, M.J., Salinas, G.L.A., Jiménez-Monreal, A.M., Chaouqi, S., Lloren, S., Martinez-Tomé, M. & Alonso, G.L. (2018) Saffron: an old medicinal plant and a potential novel functional food. *Molecules* 21(1): 30.

Barrow, J. (1749) *Dictionarium medicum universale or a new medical dictionary*. London.

Basker, D. (1999) Saffron chemistry. In: *Saffron*, Crocus sativus *L*. Ed. M. Negbi. Harwood Academic Publishers, Amsterdam. pp. 45–52.

Baziar, S., Aqamolaei, A., Khadem, E., Mortazavi, S.K., Naderi, S., Sahebolzamani, E., Mortezaei, A., Jalilevand, S., Mohammadi, M.-R., Sharmirzade, M. & Akhondzadeh, S. (2019) *Crocus sativus* L. versus methylphenidate in treatment of children with attention-deficit/hyperactivity disorder: randomized, double-blind pilot study. *Journal of Child & Adolescent Psychopharmacology* 29(3): 1–8.

Birkbeck Nevins, J. (1851) *Translation of the new London pharmacopoeia, including also the new Dublin and Edinburgh pharmacopoeias*. Longman, Brown, Green & Longmans, London.

Blacksmith, S. (2011) *The mystery of the autumn crocus* Crocus nudiflorus. Halifax Scientific Society, Halifax.

Blancard/Blanckaert, S. (1708) *The physical dictionary. 5th Ed*. London.

Boardman, T. (1805) *A dictionary of the veterinary art*. George Kearsley, London.

Bosmali, I., Ordoudi, S.A., Tsimidou, M.Z. & Madesis, P. (2017) Greek PDO saffron authentication studies using species specific molecular markets. *Food Research International* 100(1): 899 907.

Bradley, R. (1727) *The country gentleman and farmer's monthly director*. London.

Bradley, R. (1736) *The country housewife and lady's director*. London.

Brisson, A.D. (1756) *Regnum animale*. Paris.

Broadhead, G.K., Chang, A., Grigg, J. & McCluskey, P. (2016) Efficacy and safety of saffron supplementation: current clinical findings. *Critical Reviews in Food Science & Nutrition* 56(16): 2767–2776.

Brugis, T. (1640) *The marrow of physicke*. London.

Busconi, M., Colli, L., Sánchez, R.A., Santaella, M., De-Los-Mozos Pascual, M., Santana, O., Roldán, M. & Fernández, J.-A. (2015) AFLP and MS-AFLP analysis of the variation within saffron crocus (*Crocus sativus* L.) germplasm. *PLoS ONE* 10(4): e0123434. https://doi.org/10.1371/journal.pone.0123434.

Carmona, M., Zalacain, A., Pardo, J.E., López, E., Alvarruiz, A. & Alonso, G.L. (2005) Influence of different drying and aging conditions on saffron constituents. *Journal of Agricultural & Food Chemistry* 53(10): 3974–3979.

Casoria, P., Laneri, U. & Di Novella, N. (1996) A preliminary note an on interesting species of *Crocus* (*Crocus longiflorus*, Iridaceae) similar to saffron (*C. sativus*). *Economic Botany* 50(4): 463–468.

CERL website https://thesaurus.cerl.org for names of authors.

Chan, C.-C. (2010) Couching for cataract in China. *Survey of Ophthalmology* 55(4): 393–398.

Christodoulou, E., Kadoglou, N.P.E., Kostomitsopoulos, N. & Valsami, G. (2015) Saffron: a natural product with potential pharmaceutical applications. *Journal of Pharmacy & Pharmacology* 67(12): 1634–1649.

Coatsworth, Mead, Cade, Wadsworth & Hales, Drs (1718) *Pharmacopoeia pauperum or the hospital dispensatory*. London.

Cooley, A.J. (1845) *The pharmaceutical Latin grammar*. R. Groombridge & Sons, London.

Council of the Pharmaceutical Society of Great Britain (1943) *The extra pharmacopoeia. Martindale. Volume II*. Pharmaceutical Press London.

Crump, W.B. (1899) On the position of *Crocus nudiflorus* in the English flora. *The Halifax Naturalist* IV(20). (Reproduced in Blacksmith, 2011, which see).

Cullen, W. (1789) *A treatise of the materia medica. Vol. II*. Edinburgh and London.

Culpeper, N. (1683) *Pharmacopoeia Londinensis or the London dispensatory*. London.

D'Archivio, A.A., Giannitto, A., Maggi, M.A. & Ruggieri, F. (2016) Geographical classification of Italian saffron (*Crocus sativus* L.) based on chemical constituents determined by high-performance liquid-chromatography and by using linear discriminant analysis. *Food Chemistry* 212: 110–116.

Douglas, J. (1722) A botanical description of the flower and seed-vessel of the plant called *Crocus Autumnalis Sativus* that produces the true English saffron of the shops. *Philosophical Transactions* 32: 441–445.

de Tournefort, J.P. (1700) *Institutiones rei herbariae. Vol. I*. Paris.

Dhakal, S., Chao, K., Schmidt, W., Qin, J., Kim, M. & Chan, D. (2016) Evaluation of turmeric powder adulterated with metanil yellow using FT-Raman and FT-IR spectroscopy. *Food* 5(2): 36.

Dispensatorium Regium et Electorale Borusso-Brandenburgicum (1713) Berlin.

Duncan, A. (1808) *The Edinburgh new dispensatory. 4th Ed.* Ball & Bradfut, Edinburgh.

Edwards, W.P. (2000) *The science of sugar confectionery*. Royal Society of Chemistry, Cambridge.

Encyclopedie brna, s.v. Franz Matthias Hertodt [sic; should be Matthias Franz Hertodt], http://encyklopedie.brna.cz/home-mmb/?acc=profil_osobnosti&load=11938

Encyclopedie brna, s.v. Johann Phillip Hertodt, http://encyklopedie.brna.cz/home-mmb/?acc=profil_osobnosti&load=11939

Entinck, J. (1771) *The new Latin and English dictionary*. London.

Erstingium, A.C. (1770) *Nucleus totius medicinae quinque partitus oder der volkommene und allezeit fertigen Apothekers. Vol. 2*. Lemgo, Germany.

Escribano, J., Ríos, I. & Fernández, J.A. (1999) Isolation and cytotoxic properties of a novel glycoconjugate from corms of saffron plant (*Crocus sativus* L.). *Biochimica et Biophysica Acta—General Subjects* 1: 217–222.

Evelyn, J. (1699) *Acetaria: A discourse on sallets*. London.

Fernández, J.A. (2007) Genetic resources of saffron and allies (*Crocus* spp.). *Acta Horticulturae* 739: 167–185.

Finley, J.W. & Gao, S. (2017) A perspective on *Crocus sativus* L. (saffron) constituent crocin: a potent water-soluble antioxidant and potential therapy for Alzheimer's disease. *Journal of Agricultural & Food Chemistry* 65(5): 1005–1020.

Flückiger, F.A. & Hanbury, D. (1874) *Pharmacographica: A history of the principal drugs of vegetable origin*. MacMillan & Co., London.

Fontecha, J.-A. de. (1606) *Diez previlegios para mugeres prenadas. Con un diccionario medico*. Alcalá de Henares, Spain.

Francis, S. (2011) *Saffron: the story of England's red gold*. Privately published.

Francis, S.A. (in prep) Ethnobotanical uses of *Crocus* spp.

Frazer, W. (1864) *Elements of material medica. 2nd Ed.* John Churchill & Sons, London.

Furbee, B. (2009) Neurotoxic plants. In: *Clinical neurotoxicology: syndromes, substances, environments* Ed. M.R. Dobbs. Elsevier. pp. 523–542.

Gainer, J.L. (2008) *Trans*-sodium crocetinate for treating hypoxia/ischemia. *Expert Opinion on Investigational Drugs* 17: 917–924.

Gerarde, J. (1597) *The herball or generall historie of plants. Book II*. London.

Ghorbani, R. & Koocheki, A. (2017) Sustainable cultivation of saffron in Iran. *Sustainable Agriculture Reviews* 25: 169–203.

Gibson, W. (1721) *The farriers dispensatory. Part I*. London.

Gillispie, C.C. (ed.) (1970–1990) *Dictionary of Scientific Biography*, 18 vols. Scribner, New York.

General Council of Medical Education and Registration of the United Kingdom (1864) *British pharmacopoeia*. Spottiswoode & Co. Ltd., London.

General Medical Council (1899) *British pharmacopoeia 1898 4th issue*. Spottiswoode & Co. Ltd., London.

General Medical Council (1911) *British pharmacopoeia 1898 10th issue*. Spottiswoode & Co. Ltd., London.

General Medical Council (1917) *British pharmacopoeia 1914 3rd issue*. Constable & Co. Ltd., London.

Ghaffari, S. & Roshanravan, N. (2019) Saffron; An updated review on biological properties with special focus on cardiovascular effects. *Biomedicine & Pharmacotherapy* 109: 21–27.

Goodrum, M.R. (2008) Questioning thunderstones and arrowheads: The problem of recognising and interpreting stone artefacts in the seventeenth century. *Early Science & Medicine* 13: 482–508.

Graesse, J.G.T., Benedict, F., Pleschl, H. & Pleschl, S.C. (1972) *Orbis latinus: Lexikon lateinischer geographischer Namen des Mittelalters und der Neuzeit. A–D.* Klinkhardt & Bierman, Braunschweig.

Grieve, M. (1992) *A modern herbal.* Ed. C.F. Leyel Tiger Books International, London.

Grüling, P. (1665) *Medicinae practicae.* Nordhausen.

Gruner, C.E. (1855) *Homoeopathic pharmacopoeia complied by order of the German Central Union of Homoepathic Physicians.* C.H. Arnold, Leipzig.

Gurlt, H.J. & Hirsch, A. (eds.) (1884–1888) *Biographisches Lexikon der hervorragenden Ärtze aller Zeiten und Völker.* Vienna & Leipzig: Urban & Schwarzenberg. 6 vols.

Gyllenbok, J. (2018) *Encylopaedia of historical metrology, weights and measures. Vol. 3.* Birkhäuser Verlag GmbH.

Harley, J. (1876) *Royle's manual of the materia medica and therapeutics.* J. Churchill, London.

Harrison, W. (1587) Chapter VIII. Of our saffron and the dressing thereof. In: *Holinshed's chronicles of England, Scotland and Ireland, 2nd Ed., Vol. 1.* Ed. R. Holinshed. AMS Press, New York. 1965 facsimile of 1807 edition. pp. 390–394.

Harrison Martindale, W. (1932) *The extra pharmacopoeia of Martindale and Westcott. Vol. I.* H.K. Lewis & Co. Ltd., London.

Hatziagapiou, K., Lambrou, G.I. (2018) The protective role of *Crocus sativus* L. (saffron) against ischemia-reperfusion injury, hyperlipidemia and atherosclerosis: nature opposing cardiovascular diseases. *Current Cardiology Reviews* 14(4): 272–289.

Hedesan, G.D. (2016) *An alchemical quest for universal knowledge: the 'Christian Philosophy' of Jan Baptist van Helmont (1579–1644).* Routledge, London.

Heitmar, R., Brown, J. & Kyrou, I. (2019) Saffron (*Crocus sativus* L.) in ocular diseases: A narrative review of the existing evidence from clinical studies. *Nutrients* 11(3): 649.

Hempel, C.J. (1846) *Additions to the materia medica pura collected and edited by Ernst Stapf MD.* William Radde, New York; H. Balliere, London.

Hempel, C.J. (1865) *A new and comprehensive system of materia medica, 2nd Ed Vol. II.* William Radde, New York.

Hertodt, J.F. (1669) *Tartaro-Mastix Moraviae.* Vienna.

Hertodt, J.F. (1670) *Opus mirificum.* Jena.

Hertodt, J.F. (1670) Foetus croco in utero tinctus. *Miscellanea curiosa medico-physica* 1, Observatio LX, pp. 157–159.

Hertodt, J.F. (1671) *In methamorphosim ridiculam Ardensbachianam hominis pulicis.* Jena.

Hertodt, J.P. (1711) *Medicus euporistus*. The Hague.

Hertodt, M.F. (1712) *Contemplatio, super sanguinis & spiritus, hominum ad atrophiam senilem deflectentiumContemplatio, super sanguinis & spiritus, hominum ad atrophiam senilem deflectentium*. The Hague.

Hill, J. (1751) *A history of the materia medica*. London.

Holmes, C.P. (1988) *Economic activity in Saffron Walden between the sixteenth and eighteenth centuries (with particular reference to the crocus industry)*. Privately published.

Hosseini, A., Razavi, B.M. & Hosseinzadeh, H. (2018) Saffron (*Crocus sativus*) petal as a new pharmacological target: a review. *Iranian Journal of Basic Medical Science* 21(11): 1091–1099.

Hosseinzadeh, H. & Nassiri-Asl, M. (2012) Avicenna's (Ibn Sina) the Canon of Medicine and saffron (*Crocus sativus*): A review. *Phytotherapy Research* 27(4): 475–483.

Huang, W.-J., Li, F.-F., Liu, Y.-J. & Long, C.-L. (2015) Identification of *Crocus sativus* and its adulterants from Chinese markets by using DNA barcoding technique. *Iranian Journal of Biotechnology* 13(1): 36–42.

Hulsius, L. (1625) *Erste schiffart in die Orientalische Indien so die Holländische schiff im martio 1595 aussgefahren und im augusto 1597 wiederkommen verzicht. Editio quinta*. Frankfurt.

Husaini, A.M., Bhat, M.A., Kamili, A.N. & Mir, M.A. (2013) Kashmir saffron in crisis. *Current Science* 104(6): 686–687.

Hutton, C., Shaw, G. & Pearson, R. (1809) *The philosophical transactions of the Royal Society of London, from their commencement in 1665, to the year 1800*. C. & R. Baldwin, London.

Ingolfsdottir, K. & Olafsson, K. (1997) Toxicity of monkshood: a review (in Icelandic, but with English abstract). *Laeknabladid* 83(3): 163–168.

Institutiones chirurgiae, seu compendiosum Collegium Chirurgicum Universale (1737) Augsburg.

James, R. (1747) *Pharmacopoeia universalis or a new universal English dispensatory*. London.

Javadi, B., Sahebkar, A. & Emami, S.A. (2013) A survey on saffron in major Islamic Traditional Medicine books. *Iranian Journal of Basic Medical Science* 16(1): 1–11.

Johnston, W. trans. Beckman, J. (1797) *A history of inventions and discoveries. Vol. I*. J. Bell, London.

Kneschke, E.H. (1863) *Neues allgemeines deutsches Adels-Lexicon*. Friedrich Voigt, Leipzig.

Kafi, M., Kahki, A.H. & Karbasi, A. (2006) Historical background, economy, acreage, production, yield and uses. In: *Saffron* (Crocus sativus) *production and processing*. Eds. M. Kafi, A. Koocheki, M.H. Rashed & M. Nassiri. Science Publishers. pp. 1–11.

Kerrison, R.M. (1814) *An inquiry into the present state of the medical profession in Eng-*

land. Longman, Hurst, Rees, Orme & Brown, London; Laing, Edinburgh; Archer, Dublin.

Khorasanchi, Z., Shafiee, M., Kermanshahi, F., Kazaaei, M., Ryzhikov, M., Parizadeh, M.R., Kermanshahi, B., Ferns, G.A., Avan, A. & Hassainian, S.M. (2018) *Crocus sativus*, a natural food colouring and flavouring, has potent anti-tumor properties. *Phytomedicine* 43(1): 21–27.

Köhler, R. (1898) *Kleinere Schriften zur Märchenforschung.* E. Felber, Weimar.

Lach, D.F. (1977) *Asia in the making of Europe*, Vol. II, Book III, Appendix: Words of Asiatic origin introduced into the European vocabulary (sixteenth century). University of Chicago Press.

Lavaterus, L. (1596) *Of ghosts and spirites, walking by night.* London.

Leone, S., Recinalle, L., Chaivaroli, A., Orlando, G., Ferrante, C., Leporini, L., Brunetti, L. & Menghini, L. (2018) Phytotherapic use of the *Crocus sativus* L. (Saffron) and its potential applications: a brief overview. *Phytotherapy Research* 32(12): 2364–2375.

Lewis, W. (1778) *The new dispensatory. 4th Ed.* Dublin.

Lim, T.K. (2014) *Edible Medicinal and Non Medicinal Plants: Vol. 8, Flowers.* Springer. Chapter on *Crocus sativus* pp. 77–136.

London Metropolitan Archives CLC/B/192/F/001/MS11936/430/754084 Royal and Sun Insurance Group records, policy register for John Birch, 26th October 1803.

London Metropolitan Archives CLC/B/192/F/001/MS11936/472/940880 Royal and Sun Insurance Group records, policy register for John Birch, 27th April 1818.

Lopresti, A.L. & Drummond, P.D. (2014) Saffron (*Crocus sativus*) for depression: A systematic review of clinical studies and examination of underlying antidepressant mechanisms of action. *Human Psychopharmacology Clinical & Experimental* 29(6): 517–527.

Loxton, G.E. (1950) 'The true source of colchicine.' Letters to the *British Medical Journal*, Oct 14th 1950 (2): 893.

Mabberley, D.J. (2017) *Mabberley's plant-book. 4th Ed.* Cambridge University Press.

Maggi, L., Carmona, M., Kelly, S., Marigheto, N. & Alonso, G.L. (2011) Geographical origin differentiation of saffron spice (*Crocus sativus* L. stigmas)—Preliminary investigation using chemical and multi-element (H, C, N) stable isotope analysis. *Food Chemistry* 128: 543–548.

Marieschi, M., Torelli, A. & Bruni, R. (2012) Quality control of saffron (*Crocus sativus* L.): Development of SCAR markers for the detection of plant adulterants used as bulking agents *Journal of Agricultural & Food Chemistry* 60(44): 10998–11004.

Marx, W., Lane, M., Rocks, T., Ruusunen, A., Loughman, A., Lopresti, A., Marshall, S., Berk, M., Jacka, F. & Dean, O.M. (2019) Effect of saffron supplementation on symptoms of depression and anxiety: a systematic review and meta-analysis. *Nutrition Reviews* 77(8): 557–571.

Mathew, B. (1977) *Crocus sativus* and its allies (Iridaceae). *Plant Systematics & Evolution* 128: 89–103.

Mathew, B. (1999) Botany, taxonomy and cytology of *Crocus sativus* and its allies. In: *Saffron*, Crocus sativus *L.* Ed. M. Negbi. Harwood Academic Publishers, Amsterdam. pp. 19–30.

Maw, G. (1886) *A monograph of the genus* Crocus, *with an appendix on the etymology of the words crocus and saffron by C.C. Lacaita.* Dulau & Co., London.

Mayne, R.G. (1860) *An expository lexicon of the terms, ancient and modern, in medical and general science.* John Churchill, London.

McClymont, J.R. (1920) *Essays on early ornithology and kindred subjects.* Bernard Quartich Ltd., London.

McGee, H. (2004) *McGee on food and cooking: an encyclopedia of kitchen science, history and culture.* Hodder & Stoughton, London.

Melnyk, J.P., Wang, S. & Marcone, M.F. (2010) Chemical and biological properties of the world's most expensive spice: saffron. *Food Research International* 43(8): 1981–1989.

Miller, J. (1722) *Botanicum officinale, or a compendious herbal.* London.

Miller, P. (1759) *The gardeners dictionary. 7th Ed.* London.

Mojon, B. (1783) *Pharmacopoea manualis reformata.* Genoa.

Molina, R.V., García-Luis, A., Coll, V., Ferrer, C., Valero, M., Navarro, Y. & Guardiola, J.L. (2004) Flower formation in the saffron crocus (*Crocus sativus* L.). The role of temperature. In: *Proceedings of the 1st International Symposium on Saffron.* Eds. J.-A. Fernández & F. Abdullaev. *Acta Horticulturae* 650: 39–47.

Mollazadeh, H., Emamai, S.A. & Hosseinzadeh, H. (2015) Razi's Al-Hawi and saffron (*Crocus sativus*): a review. *Iranian Journal of Basic Medical Science* 18(12): 1153–1166.

Moradi-Khatoonabadi, Z. Amirpour, M. & AkbariAzam, M. (2015) Synthetic food colours in saffron solutions, saffron rice and saffron chicken from restaurants in Tehran, Iran. *Food Additives & Contaminants: Part B* 8(1): 12–17.

Moradzadeh, M., Kalani, M.R. & Avan, A. (2019) The antileukemic effects of saffron (*Crocus sativus* L.) and its related molecular targets: a mini review. *Journal of Cellular Biochemistry* 120(4): 4732–4738.

Moratalla-López, N., Zalacain, A., Bagur, M.J., Salinas, M.R. & Alonso, G.L. (2019a) Saffron. In: *Food integrity handbook. a guide to food authenticity issues and analytical solutions.* Eds J.-F. Morin & M. Lees. https://secure.fera.defra.gov.uk/foodintegrity/secure/downloadFile.cfm?id=614 accessed on 17/01/2019.

Moratella-López, N., Bagur, M.J., Lorenzo, C., Martinez-Navarro, M.E., Salinas, M.R. & Alonso, G.L. (2019b) Bioactivity and bioavailability of the major metabolites of *Crocus sativus* L. flower. *Molecules* 24(15): 2827.

Mortimer, T. (1810) *A general dictionary of commerce, trade and manufactures.* Richard Phillips, London.

Mosan, J. trans. Wirtzung, C. (1605) *The general practise of physicke*. London.

Motherby, G. & Wallis, G. (1745) *A new medical dictionary, or general repository of physic. 4th Ed.* London.

Murray, J. (1824) *A system of materia medica and pharmacy*. Duyckinck, Long, Collins & Co., New York.

Naeimi, M., Shafiee, M., Kermanshahi, F., Khorasanchi, Z., Khazaei, M., Ryzhikov, M., Avan, A., Gorji, N. & Hassanian, S.M. (2019) Saffron (*Crocus sativus*) in the treatment of gastrointestinal cancers: current findings and potential mechanisms of action. *Journal of Cellular Biochemistry* 120(10): 16330–16339.

National Convention for Revising the Pharmacopoeia (1893) *The pharmacopoeia of the United States of America. Seventh decennial revision*. P. Blakiston & Son, Philadelphia.

Needham, J. (1959) *A history of embryology*. Cambridge University Press.

Negbi, M. (1999) Saffron cultivation: past, present and future prospects. In: *Saffron*, Crocus sativus *L*. Ed. M. Negbi. Harwood Academic Publishers, Amsterdam. pp. 1–18.

Nemati, Z., Blattner, F.R., Kerndorff, H., Erol, O. & Harpke, D. (2018) Phylogeny of the saffron-crocus group, Crocus series Crocus (Iridaceae). *Molecular Phylogenetics & Evolution* 127: 891–897.

Nemati, Z., Harpke, D., Gemicioglu, A., Kerndorff, H. & Blattner, F.R. (2019) Saffron (*Crocus sativus*) is an autotriploid that evolved in Attica (Greece) from wild *Crocus cartwrightianus. Molecular Phylogenetics & Evolution* 136: 14–20.

Newman, L.F. (1945) Some notes on folk medicine in the Eastern Counties. *Folklore* 56(4): 349–360.

Ng, L.T., Chang, Y.S. & Kadir, A.A. (1997) A review on agar (gaharu) producing *Aquilaria* species. *Journal of Tropical Forest Products* 2(2): 272–285.

Norri, J. (2016) *Dictionary of medical vocabulary in English, 1375–1550: Body parts, sicknesses, instruments and medicinal preparations*. Routledge.

Ordoudi, S.A., Cagliani, L.R., Melidou, D., Tsimidou, M.Z. & Consonni, R. (2017) Uncovering a challenging case of adulterated commercial saffron. *Food Control* 81: 147–155.

Oswald, P.H. & Preston, C.D. (2011) *John Ray's Cambridge catalogue (1660)*. The Ray Society, London.

Parkinson, J. (1629) *Paradisi in sole paradisus terrestris*. London.

Parkinson, J. (1640) *Theatrum botanicum*. London.

Parr, B. (1819) *The London medical dictionary Vol. 1*. Mitchell, Ames & White, Philadelphia.

Partridge, J. trans. Mynsicht, H. (1682) *Thesaurus & armamentarium medico-chymicum, or, a treasury of physic*. London.

Patel, S., Sarwat, M. & Khan, T.H. (2017) Mechanism behind the anti-tumour potential of saffron (*Crocus sativus* L.): the molecular perspective. *Critical Reviews in Oncology/Hematology* 115: 27–35.

Pechey, J. (1705) *The compleat herbal of physical plants*. London.

Pereira, J. (1840) *The elements of the material medica. Part II*. Longman, Orme, Brown, Green & Longmans, London.

Pharmaceutical Society of Great Britain (1907) *The British pharmaceutical codex*. Pharmaceutical Society, London.

Pharmacopoea Wirtenbergica (1754) Stuttgart.

Pharmacopoeia Augustana renovata et aucta (1684) Augsburg.

Philanthropos, T. (1741) *A physcial vade mecum*. London.

Pieroni, A., Zaman, H., Ayub, S. & Torry, B. (2010) 'My doctor doesn't understand why I use them': Herbal and food medicines amongst the Bangladeshi community in West Yorkshire, UK. In: *Ethnobotany of the new Europe*. Eds. M. Pardo-de Santayana, A. Pieroni & R.K. Puri. Bergham Books, Oxford.

Plant Breeding Matters, Spring 2012 (information from the British Society of Plant Breeders), accessed from http://www.wheatisp.org/PDFs/BSPB_Plant_Breeding_Matters.pdf

Pourmasoumi, M., Hadi, A., Najafgholizadeh, A., Kafeshani, M. & Sahebkar, A. (2019) Clinical evidence on the effects of saffron (*Crocus sativus* L.) on cardiovascular risk factors: a systematic review meta-analysis. *Pharmacology Research* 139: 348–359.

Pullela, R., Young, L., Gallagher, B., Avis, S.P. & Randell, E.W. (2008) A case of fatal aconitine poisoning by monkshood ingestion. *Journal of Forensic Science* 53(2): 491–494.

Quattrocchi, U. (2016) *CRC world dictionary of medicinal and poisonous plants*. CRC Press, Boca Raton, USA.

Quincy, J. (1719) *Pharmacopœia officinalis & extemporanea. Or, a complete English dispensatory. 2nd Ed.* London.

Quincy, J. (1733) *Pharmacopœia officinalis & extemporanea. Or, a complete English dispensatory. 9th Ed.* London.

Rahmani, J., Manzari, N., Thompson, J., Clark, C.C.T., Villaneuva, G., Varkaneh, H.K. & Mirmiran, P. (2019) The effect of saffron on weight and lipid profile: a systemic review, meta-analysis, and dose-response of randomised clinical trials. *Phytotherapy Research* 33(9): 2244–2255.

Raja, A.S.M., Pareek, P.K., Shakyawar, D.B., Wani, S.A., Nehvi, F.A. & Sofi, A.H. (2012) Extraction of natural dye from saffron flower waste and its application on pashmina fabric. *Advances in Applied Science Research* 3(1): 156–161.

RASFF (EU's Rapid Alert System for Food and Feed) https://webgate.ec.europa.eu/rasff-window/portal/?event=SearchByKeyword&NewSearch=1&Keywords=saffron accessed 11/01/2019.

Redwood, T. (1847) *Gray's supplement to the pharmacopoeia; being a concise but comprehensive dispensatory and manual of facts and formulae for the chemist, druggist and medical practitioner*. Longman & Co.; Simpkin & Co.; Sherwood & Co.; John Churchill; Henry Bohn; and Henry Renshaw: London.

Register of the local families of the town of Nikolsburg, (http://www.nikolsburg.eu/OFB/Nikolsburg/15/ofb1k35973.html, accessed 11/09/17).

Rekha, K., Koul, S. & Ram, G. (2011) Adulteration and marketing of saffron (*Crocus sativus* Linn.). *Journal of Tropical Medicinal Plants* 12(1): 135–140.

Rubert, J., Lacina, O., Zachariasova, M. & Haslova, J. (2016) Saffron authentication based on liquid chromatography high resolution tandem mass spectrometry and multivariate data analysis. *Food Chemistry* 204: 201–209.

Rukšāns, J. (2010) *Crocuses: a complete guide to the genus.* Timber Press, Portland (Oregon) & London.

Rukšāns, J. (2017) *The world of crocuses.* Latvian Academy of Sciences, Riga.

Sajner, J. (1974) Johann Ferdinand von Todtenfeld. In: *Proceedings of the XXIII International Congress of the History of Medicine, London, 2–9 September 1972.* London: Wellcome Institute for the History of Medicine, pp. 1137–1139.

Salmon, W. (1694) *Pharmacopoeia Bateana or Bate's dispensatory.* London.

Schenck, J.G. (1607) *Pandectae sive partitiones medicinales.* Frankfurt.

Schmidt, T., Heitkam, T., Liedtke, S., Schubert, V. & Menzel, G. (2019) Adding colour to a century-old enigma: multi-colour chromosome identification unravels the autotriploid nature of saffron (*Crocus sativus*) as a hybrid of wild *Crocus cartwrightianus* cytotypes. *New Phytologist* 222(4): 1965–1980.

Schröder, J. (1685) *Trefflich versehene medicin-chymische apotheke.* Nuremberg.

Seidermann, J. (2005) *World spice plants: economic usage, botany, taxonomy.* Springer-Verlag, Berlin, Heidelberg & New York.

Serrano-Díaz, J., Sánchez, A.M., Martínez-Tomé, M., Winterhalter, P. & Alonso, G.L. (2013) A contribution to nutritional studies on *Crocus sativus* flowers and their value as food. *Journal of Food Composition and Analysis* 31(1): 101–108.

Shakiba, M., Moazen-Zadeh, E., Noorbala, A.A., Jafarinia, M., Divsalar, P., Kashani, L., Shahmansouri, N., Tafakhori, A., Bayat, H. & Akhondzadeh, S. (2018) Saffron (*Crocus sativus*) versus duloxetine for treatment of patients with fibromyalgia: a randomized double-blind clinical trial. *Avicenna Journal of Phytomedicine* 8(6): 513–523.

Shuttleworth, E.B. (1876) Adulteration of saffron (editorial). *Canadian Pharmaceutical Journal* 9: 368–369.

Sonnedecker, G. (1963) *Kremer's and Urdang's history of pharmacy. 3rd Ed.* J.B. Lippincott & Co., Philadelphia & Montreal.

Smedley, E., Cooke Taylor, W., Thompson, H. & Rich, E. (1855) *The occult sciences: Sketches of the traditions and superstitions of past times and the marvels of the present day.* Richard Griffin & Co., London & Glasgow.

Stother, E. (1727) *Materia medica or a new description of the virtues and effects of all the drugs and simple medicines now in use.* London.

Strybel, R. & Strybel, M. (2005) *Polish heritage cookery.* Hippocrene Books, New York.

Sun, J., Freeman, B.D. & Natanson, C. (2018) Meta-analysis of clinical trials. In: *Principles and practice of clinical research. 4th ed.* Eds. J.I. Gallin, F.P. Ognibene & L.L. Johnson. Academic Press. pp. 317–327.

Sweringen, H.V. (1873) *Pharmaceutical lexicon.* Laindsay & Blakiston, Philadeliphia.

Tammaro, F. (1999) Saffron (*Crocus sativus* L.) in Italy. In: *Saffron,* Crocus sativus *L.* Ed. M. Negbi. Harwood Academic Publishers, Amsterdam. pp. 53–62.

The Plant List v. 1.1 (www.theplantlist.org)

Thiercelin, J. (2003) *Saffron: the gold of cuisine.* Agnès Viénot Éditions, Paris.

Tomlinson, R. trans. de Renou, J. (1657) *A medical dispensatory, containing the whole body of physick.* London.

Tong, Y., Zhu, X., Yan, Y., Liu, R., Gong, F., Zhang, L., Hu, J., Fang, L., Wang, R. & Wang, P. (2015a) The influence of different drying methods on constituents and antioxidant activity of saffron from China. *International Journal of Analytical Chemistry* 2015. 953164. doi:10.1155/2015/953164

Tong, Y., Yan, Y., Zhu, X., Liu, R., Gong, F., Zhang, L. & Wang, P. (2015b) Simultaneous quantification of crocetin esters and picrocrocin changes in Chinese saffron by high-performance liquid chromatography-diode array detector during 15 years of storage. *Pharmacognosy Magazine* 11(43): 540–545.

Tóth, B., Hegyi, P., Szakács, Z., Kerémi, B., Varga, G., Pétervári, E., Balaskó, M., Rumbus, Z., Rakonczay, Z., Bálint, E.R., Kiss, T. & Csupor, D. (2019) The efficacy of saffron in the treatment of mild to moderate depression: a meta-analysis. *Planta Medica* 85(1): 24–31.

Uddin, M.S., Al Mamun, A., Kabir, M.T., Jakaria, M., Mathew, B., Barreto, G.E. & Ashraf, G.M. (2019) Nootropic and anti-Alzheimer's actions of medicinal plants: molecular insight into therapeutic potential to alleviate Alzheimer's neuropathology. *Molecular Neurobiology* 65(7): 4925–4944.

United States Pharmacopoeial Convention (1907) *The pharmacopoeia of the United States of America. Eighth decennial revision.* P. Blakiston, Son & Co., Philadelphia.

US Patent US6060511A. *Trans*-sodium crocetinate, methods of making and methods of use thereof.

Vavilov, N.I. (1992) *Origin and geography of cultivated plants.* Cambridge University Press.

Walton, I. (1653) *The compleat angler.* London.

Waring, E.J. (1878) *Bibliotheca therapeutica. Vol. 1.* The New Sydenham Society, London.

Wibmer, K. (1831) *Die wirkung der arzneimittel und gifte im gersunden thierischen körper. 1. Band.* Literarisch artistischen Anstalt, Munich.

Wieand, P.R. (1961) *Folk medicine plants used in the Pennsylvania Dutch country.* Wieand's Pennyslvania Dutch, USA.

Williard, P. (1998) *Secrets of saffron.* Souvenir Press, London.

Wollkopf, J.B. (1811) *Pharmacopoea extemporanea oder handbuch für selbstdispensirende Ärzte am krankenbette.* Johann Ambrosius Barth, Leipzig.

Wood, G.R. & Bache, F. (1865) *Dispensatory of the United States of America. 12th Ed.* Philadelphia.

Wood, C.H. & Sharp, C. (1866) *The year-book of pharmacy* (for 1865). John Churchill & Sons, London.

World Register of Marine Species http://www.marinespecies.org/aphia.php?p=taxdetails&id=135124#distributions (accessed on 12/9/2018)

Xiao, W., Li, S., Wang, S. & Ho, C.-T. (2017) Chemistry and bioactivity of *Gardenia jasminoides*. *Journal of Food & Drug Analysis* 25(1): 43–61.

Yaribegi, H., Zare, V., Butler, A.E., Barreto, G.E. & Sahebkar, A. (2018) Anti-diabetic potential of saffron and its active constituents. *Journal of Cellular Biology* 234(6): 8610–8617.

Zampieri, F. & Zanatta, A. (2015) Medical history in Italy: the University of Padua medical school. In: *Medical humanities: Italian perspectives* (Eds. Caenazzo, L., Mariani, L. & Pegorarao, R.). CLEUP, Padua. pp. 31–38.

Zedler, J.H. (1793) *Grosses vollständiges universal lexicon. Vol. 6*. Halle & Leipzig.

Zeka, K., Ruparelia, K.C., Conintenza, M.A., Stagos, D., Vegliò, F. & Arroo, R.R.J. (2015) Petals of *Crocus sativus* L. as a potential source of the antioxidants crocin and kaempferol. *Fitoterapia* 107: 128–134.

Zohary, D. & Hopf, M. (2000) *Domestication of plants in the Old World. 3rd Ed.* Oxford University Press.

Index

To keep the size of this index manageable, focus has been placed on diseases, treatments, formulae, materia medica, saffron and the many authors cited by Johann Ferdinand Hertodt. For more information on the materia medica, see Appendix 3. For further details on the Hertodt's references, see Appendix 1.